TELECOMMUNICATIONS CONVERGENCE

McGraw-Hill Telecommunications

TELECOMMUNICATIONS CONVERGENCE

BRIDGING THE GAP BETWEEN TECHNOLOGIES AND SERVICES

STEVEN SHEPARD

McGraw-Hill
New York   Chicago   San Francisco
Lisbon   London   Madrid   Mexico City   Milan
New Delhi   San Juan   Seoul   Singapore
Sydney   Toronto

338.4762/382
554t2

Cataloging-in-Publication Data is on file with the Library of Congress.

McGraw-Hill

A Division of The McGraw·Hill Companies

Copyright © 2002 by The McGraw-Hill Companies, Inc. All rights reserved.
Printed in the United States of America. Except as permitted under the United
States Copyright Act of 1976, no part of this publication may be reproduced or distributed in any form or by any means, or stored in a database or retrieval system,
without the prior written permission of the publisher.

1 2 3 4 5 6 7 8 9 0   DOC/DOC   0 7 6 5 4 3 2

ISBN 0-07-138785-4

The sponsoring editor for this book was Steven Chapman and the production
supervisor was Pamela Pelton. It was set in Fairfield by MacAllister Publishing
Services, LLC.

Printed by R. R. Donnelley & Sons Company.

McGraw-Hill books are available at special quantity discounts to use as premiums
and sales promotions, or for use in corporate training programs. For more information, please write to the Director of Special Sales, McGraw-Hill, 2 Penn Plaza,
New York, NY 10121-2298. Or contact your local bookstore.

This book is printed on recycled, acid-free paper contained a minimum of
50 percent recylced de-inked fiber.

For my Dad. You always said, "Do it because you love it."
Now I understand.

ABOUT THE AUTHOR

Steven Shepard is a professional writer and educator who specializes in international telecommunications. Formerly with Hill Associates, he is also the author of *SONET/SDH Demystified*, *Telecom Crash Course*, and *Optical Networking Crash Course*. He lives and works in Williston, Vermont.

CONTENTS

ACKNOWLEDGMENTS

This book has benefited greatly from the efforts of many. I extend heartfelt thanks to Cyril Berg, Rich Campbell, Joe Cappetta, Joe Carlisle, Phil Cashia, Bob Dean, Mark Fei, Jack Garrett, Jack Gerrish, Jerry Hanley, Dave Hill, Steve Hillier, Barbara Jorge, Gary Kessler, Phyllis Klees, Sergei Kuharsky, Naresh Lakhanpal, Joe Lazzara, Al Lounsbury, Gary Martin, Mitch Moore, Richie Parlato, Barbara Potter, George Powch, Rick Sanders, Kenn Sato, Kirk Shamberger, Henry Sherwood, Dave Stubbs, Elvia Szymanski, Jim Taylor, Christine Troianello, Dave Whitmore, and Robin-Marie Williams.

Steve Chapman at McGraw-Hill has been in my corner from the beginning, and I owe him much. Thanks, Steve, for continuing to believe in the project and for the long, fascinating (sometimes even relevant!) conversations. Jessica Hornick, also of McGraw-Hill, kept me on track and provided the kind of prodding, support, and perspective that only an author can appreciate.

I extend a special thank you to George Soscia of St. Louis, Missouri. Shortly after the first edition of the book appeared on shelves, George appeared in my e-mail as a voluntary, unsolicited editor who contributed countless hours to the task of finding typos and errors as he read the book and sending them to me. The volume of them was embarrassing; the value of them was priceless. Thank you, George.

As always, very special thanks to my wife, Sabine, and my kids, Steven and Cristina. I hope I've been there for you as much as you have always been there for me. I love you.

FOREWORD

Too often the fascinating possibilities of technological innovation have caused the telecommunications industry's leaders to turn their attention away from the needs of their customers and sent them rushing down the dangerous path of deploying technology for technology's sake. Inevitably, industry contraction and confusion have followed, much to the detriment of not just the industry, but those who stood to benefit most from the applications of the technology.

Today the most articulate advocate for the importance of maintaining a focus on the industry's customers is Steve Shepard, author of *Telecommunications Convergence* and numerous other works that help explain where the telecom industry is heading and why. His message (that the inexorable convergence of technology, services, and corporate structures must form the foundation of any successful strategy in this rapidly changing industry) has never been more powerfully articulated than in this revised and updated edition. While the rest of the world tries to make sense out of the recent collapse of the .com bubble, much of it fueled by over investments in telecommunication technologies and related business schemes, Steve Shepard presents a cogent analysis of what went wrong and why the industry still stands on the brink of unprecedented success in the future. His message is at the heart of the Center for Telecommunications Management executive management classes because it presents the best prescription for an industry recovery that can spur a new round of global economic growth.

Readers of this completely revised edition will gain a comprehensive and unique understanding of the telecommunications industry's future that they can use to be successful with their own business strategies. The Center for Telecom Management at USC's Marshal School of Business strongly endorses this book and hopes its message will help make it possible for customers everywhere to enjoy the benefits of this powerful technology now and in the future.

Morley Winograd
Executive Director
Center for Telcom Management
Marshall School of Business at USC

PREFACE

WHAT A DIFFERENCE A YEAR AND A HALF MAKES

The first edition of this book arrived in bookstores in May 2000. It took a different approach than most technical works, concluding that although telecommunications technologies are centrally important to the success of (1) the telecommunications industry itself and (2) the companies that benefit from its technology-based products, it is the services that those technologies make possible that bring the true value and promise of telecommunications. In the book, I argued that the business philosophies of the long-time players in the industry had to change if the companies were to survive, much less flourish, in the evolving knowledge-driven economy. I cautioned companies that saw themselves as pure technology providers that commodities confer short-lived marketplace advantages and that the technologies they sold tended to commoditize rapidly. Success, I argued, depended upon the ability to sell packaged solutions that address real business problems for the customers as well as the customers' customers.

There is nothing wrong, of course, with the commodity business—commodities are commodities, after all, because large numbers of people want them. The darker side of that, however, is that commodities by classical definition are products that can only be differentiated by price. Companies that can aggregate them, create a market, and sell in huge volumes at the right time can make a healthy profit, as long as they never lose sight of the business that they are *really* in and the product that they *really* sell. The business is solutions delivery; the product is competitive advantage for the customer. The fact that the delivery is facilitated by effectively deployed technology is an important detail, but a detail nonetheless.

For some companies, this is a bitter pill to swallow. For years telecomm players have simmered in the juices of technology, taking on its essence and, as my Zen friends would say, becoming one with it. Many have come to believe that the delivery of technology is an end unto itself, and for some, it is. In the last two years, however, the *service providers* have been forced to confront a very uncomfortable fact: that dial tone, access, and transport are not services. They are commodities, offered by a growing number of players, and are purchased by customers who expect them to be, for all intents and purposes, free. In fact, recent regulatory decisions designed to create an artificially competitive market have conveyed this perception to the market, a perception that is at best wrong and at worst highly damaging to the industry.

This is unfortunate. Today's service providers have invested enormous amounts of money into their access and transport infrastructures, initially doing so as part of a large service monopoly. The advantage that this conveys is one of presence and capability. The disadvantage is that the network is in many cases based on older, less flexible technology that is incapable of adapting to the rigors of modern demands for service. Some of the incumbent players have upgraded their in-place infrastructure, but at significant cost: Those same regulatory decisions mentioned earlier mandate long depreciation periods for infrastructure that at today's accelerated pace of technological change becomes functionally obsolete well before the asset is depreciated. However, the challenge is more fundamental than that: Even if an incumbent player has the financial or technological wherewithal to invest in new technology, the company will often not pursue the investment because under current regulatory strictures, any newly deployed technology must be immediately made available to the incumbent's competitors, thus eliminating any first mover advantage that might have been enjoyed by the incumbent provider. Solutions to this conundrum do exist, but they are elusive and run contrary to the reigning philosophy of the industry.

In the first edition of this book, I observed that the convergence phenomenon is real, necessary, and comprised of three pieces: technology convergence, which described the apparently inexorable evolution to a single technology-agnostic con-verged network capable of delivering a wide array of sellable ser-vices, company convergence, which describes the tendency of companies to coalesce as a way of consolidating capability, and services convergence, the use of knowledge management tech-niques by converged companies to develop profiles of clients so that solutions can be built that address their particular business challenges.

In technology convergence, we discussed the transport, access, and premises technologies that make up the modern telecommunications network. Within the transport arena we examined point-to-point and switched options including T-1, *asynchronous transfer mode* (ATM), *frame relay* (FR), *Internet protocol* (IP) solutions, and integrated combinations of these. Under access, we explored traditional twisted pair, *Integrated Services Digital Network* (ISDN), the many flavors of *digital subscriber line* (DSL), cable-based access technologies, and both fixed and mobile wireless. Finally, we examined premises-based options, including *local area networks* (LANs) and the solutions proposed by the *Home Phoneline Networking Alliance* (HomePNA).

As the book moved into company convergence, the scene shifted from technology to the evolution of business. We dis-cussed the shift in focus from e-commerce to e-business and examined the four most common corporate models found in the e-economy: the flea market, the arbitrator, the process improver, and the virtual corporation. We discussed the enter-prise resource planning process and the manner in which it is used by the various segments of the telecommunications indus-try, concluding with an analysis of service providers and manu-facturers and the degree to which they fall into one of the four corporate categories.

Finally, in services convergence, we pulled it all together. Technology convergence is a facilitator, company convergence is a convenience mechanism, and services convergence is the ultimate goal—the distinguished deliverable that provides dif-ferentiation among product offerings that would otherwise be relegated to commodity status. In this final section we dis-cussed the critical role of measurable, sustainable *quality of service* (QoS) of the signaling and packaging protocols, such as *multiprotocol label switching* (MPLS) and *Media Gateway Control Protocol* (MGCP), that make QoS possible and the services that are emerging in the knowledge-driven telecommunications world. We discussed call centers, IP-enhanced *private branch exchanges* (PBXs), and the advantages and disadvantages of private (read corporate) versus public network infrastructures.

Now, 18 months later, as I prepare to write the second edition of the book, I am finding that a revision and update just won't do it. For most technical books, a revision of standards and products and a look at the companies in the industry segment are usually enough to create the second edition of the book. That's not the case here. Although the technologies still work the same way they did when the first edition was released, the relationships between them and the degree of their acceptance and utilization are dramatically different. MPLS has emerged as a clear winner in the QoS control arena and is far more widely deployed than it was a year and a half ago. IP continues to gain ground, and services based on it continue to grow, but they have not become the world changers that many predicted. FR is still out there, but its future has become a bit murky and still sits in ATM's shadow. ATM continues to find a home in *wide area networks* (WANs) and continues to be the best current hope for QoS delivery in switched networks. Together with IP, often with MPLS in the middle, ATM provides a powerful solution that will continue to be deployed for some time to come.

Other changes have occurred as well. DSL in all its glory faces an uncertain future because of distance challenges, deployment difficulties, installation failures, and competition from cable modems and fixed wireless access. Wireless, widely seen as a technological savior for broadband, has proven to be less than capable of succeeding in the marketplace, but not for technical reasons—it does precisely what it was designed to do —the failure comes from service provider tendencies to over-promise its capabilities, resulting in massive market disappointment. Cable has enjoyed enormous market uptake, the result of which has been enormous market disappointment as the technology has succumbed to its own success. As I write this, my own cable modem sits idle due to massive overuse by subscribers who share the network's bandwidth. Meanwhile, ISDN enjoys a renaissance of sorts in some markets due to the failures of DSL and cable to live up to their considerable capability promises. Ethernet, once seen as an in-building solution for office data transport, has emerged as a powerful contender in metro networking and is even being touted as a low-cost, high-bandwidth framing solution for long-haul and metro providers.

And what of the companies discussed in the company convergence section of the first edition? Well, to a certain extent, one way or another they definitely converged. Of the 170-odd corporations used as examples of convergence (both successful and not), more than 30 have disappeared, 20 have been acquired, and *all* have suffered the whiplash of the current economic downturn caused by the mania of investment in the telecommunications market and exacerbated by the September 11 attacks. Interestingly, when I examine the list, the following facts jump out at me:

- Pure technology providers (DSL and wireless access) were the first to go.

- Incumbent service providers that own expansive networks weathered the storm reasonably well, with significant capital investment plans underway.

- Competitors to the incumbents, particularly *competitive local exchange carriers* (CLECs) and *Internet service providers* (ISPs), have for the most part not done well for a variety of reasons that will be explained later in the book.

- Online *services* providers, including *application service providers* (ASPs), professional services firms, and content providers, have been minimally affected by the economic downturn.

- Hardware manufacturers with broad, diffuse product lines have retrenched and been forced to reconsider lines of business and corporate intent.

Service convergence, the ultimate goal of the three-part convergence phenomenon, has its own set of observations. Measurable and sustainable QoS became the deliverable, and services based on customer requirements rather than technological capabilities became the chanted mantra of the industry. The goal, however, has been elusive. Although some companies have succeeded in focusing on and delivering services based on stated customer requirements, others have tried, but have fallen back on deliverables that they are comfortable with—for the most part, technology. Although technology certainly plays a central role in the ability to deliver customer service, it is really nothing more than a facilitator of that service. Services must be defined by the customer, not by the service provider; technologies, on the other hand, are typically defined and offered by the service provider as a solution in their own right. More often than not, however, technologies positioned as complete solutions are not successful because they fail to take into account concerns and issues that technology alone cannot address.

In the early 1960s, American psychologist Abraham Maslow proposed his now-famous five-layer hierarchy of basic human needs. In his hierarchy, Maslow ranked the basic human needs as follows: physiological demands, security and safety concerns, love and feelings of belonging, competence, prestige, and esteem, self-fulfillment, and finally, curiosity and the need to understand the surrounding world. Maslow concluded that the basic requirements had to be fulfilled before the higher level, more cognitive demands could be fulfilled.

Interestingly, there is a similarly structured hierarchy of needs in the domain of telecommunications, particularly as it relates to the activities of network designers, service providers, and manufacturers of components and network devices. This hierarchy, shown in Figure I, illustrates a seven-layer hierarchy: features, functions, benefits, applications, services, solutions, and finally, value. All of these words appear in various marketing, sales, and technical documentation, and all are important. However, the manner in which they are used and the qualities that they each represent deserve attention.

The *American Heritage Dictionary* defines the seven terms as follows:

FIGURE I   The hierarchy of value.

Feature   A prominent or distinctive aspect, quality, or characteristic

Function   The action for which (something) is particularly fitted or employed

Benefit   Something that promotes or enhances well-being; an advantage

Application   The act of putting something to a special use or purpose

Service   An act of assistance or benefit to another or others

Solution   The method or process of solving a problem

Value   Worth in usefulness or importance to the possessor; utility or merit

There are two ways to approach the use of this network services hierarchy. One way is from the top down, starting with features and passing through the various layers of the model; the other is to start with value at the bottom and work upward. Both are valid, but must be utilized appropriately. Let's consider the functional differences between the seven layers.

Features, functions, and benefits are characteristics that define the technical capabilities and defining parameters of a device or service. A prominent or distinctive aspect, quality, or characteristic, the definition of a feature, clearly speaks to such things as the physical footprint (space a device occupies in a central office), the amount of heat it generates while in operation, the amount of electricity it consumes, backplane capacity, and component redundancy. Function, the action for which (something) is particularly fitted or employed, describes the technical inner workings of a device or software module that result in some sort of operational value for the purchaser of the product. A benefit, something that promotes or enhances well-being; an advantage, provides precisely that: an advantage that the product conveys to the customer, although typically interpreted as a technical rather than a market advantage, such as the ability to perform a hot swap of a component.

Moving down the stack, we come to application, the act of putting something to a special use or purpose. An application refers to the manner in which a product is actually used—usually the reason that the customer purchased the item in the first place and the first occurrence of possible value to the consumer. Service, an act of assistance or benefit to another or others, refers to the process of converting what is usually a generic application set into a more focused, almost customized treatment to address a specific set of needs for a client. Solutions, defined as methods or processes used to resolve a problem, carry the specificity of services to the next level, addressing the customer's specific business requirements and often presenting a product directed as much at the customer's customer as at the actual customer.

Finally, we come to value, something that is useful or important to the possessor and offers utility or merit. Value is the most critical of the seven elements because it is universal and timeless. Value, custom-defined in the mind of each customer, is a personal, specific, and difficult-to-quantify essence, unique to every client. The other six—feature, function, benefit, application, service, and solution—are relatively static and time-dependent. In other words, an application that is timely, useful, and effective today may not be six months from now. A solution that resolves today's vexing customer problem may not do so this time next year. Value, however, is a constant and has little to do with technology. It has everything to do, however, with a discrete knowledge and understanding of what makes the customer tick.

There are a number of other analogies that lend themselves to this discussion. Consider for a moment the *Open Systems Interconnection* (OSI) reference model. Its seven protocol layers cover the waterfront from physical transmission of bits to the lyrical interpretation of the meaning of specific data structures at the application layer. However, there is another hierarchy at work here that is often overlooked, the hierarchy of customer proximity.

The lower three layers of the OSI model—physical, data link, and network—have the following characteristics. They are based on clearly defined, widely accepted international standards, there is relatively little room for interpreting their very clearly stated intentions, they rarely change, they are deeply embedded in the core of the network itself, and with few exceptions, they rarely touch the customer. Finally, they represent the current operational domain of the typical service provider. Transport, switching, and routing are the responsibilities of the telephone companies, cable companies, and their various transport cousins.

The upper layers (transport through application) have varied responsibilities and are characteristically quite different from the lower layers. They are open to broad interpretation because they are close to the customer and must therefore be able to accommodate the diverse requirements of diverse application types; although based on standards, the standards are fluid and constantly being augmented or modified to meet the changing demands of the clients they serve. They are found scattered around the periphery of the network because that's where the clients are, and finally, they represent the operational domain of content providers, service providers (in the truest sense of the term), ASPs, and other companies that rely on the underlying network to transport their traffic, highly customized for each customer. Clearly, then, there is a functional separation that occurs between the bottom and top of the OSI food chain. At the bottom, the network reigns supreme; at the top, the customer is emperor of all that he or she surveys.

In the days when bandwidth held sway and was a tremendous money generator, the lower layers represented a cash cow ripe for milking. Today, however, with the perceived glut of bandwidth that is now available, the price of bandwidth has plummeted to near-zero, a frightening reality for those companies that have traditionally made their fortunes through the sale of bits-per-second. Today, the big money lies at the top of the stack, closer to the customer, a place where unique, specially designed network products can be positioned on a customer-by-customer basis. The traditional service providers—the *incumbent local exchange carriers* (ILECs), CLECs, and *interexchange carriers* (IXCs)—are scrambling to establish a toehold that will enable them to climb the stack, to move out of the primordial network ooze into the lofty heights of content and services. Of course, they are in an enviable position: They touch the customers with their networks, and whoever is closest to the customer has the greatest influence over them and the infrastructure deployment decisions they are likely to make. The combination of network provisioning and content is unbeatable and is fast becoming a major focus for converged providers. The ability to provision an end-to-end network as well as deliver application content is a powerful combination. Thus, the service providers' interest in moving up the food chain is a valid one. They do have a challenge before them, however, illustrated in Figure II. The illustration shows two triangles. One is labeled Knowledge, the other, $. The Knowledge pyramid has a broad base, while the $ pyramid has a broad top. Above the pyramids we find the upper layers, below, the lower layers. The significance of this graphic is that the traditional service providers enjoy a broad base of knowledge about the lower three layers, given the fact that they have been designing, building, and maintaining them for more than 125 years. Furthermore, it costs them very little to operate there because of their massive embedded base.

On the other hand, they have precious little knowledge about the upper layers, and the accumulation of that knowledge

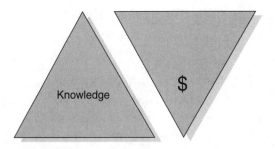

FIGURE II   Knowledge versus cost.

is an extremely expensive undertaking, albeit a necessary one. The process of adding capability within the higher layer services to their collection of existing skills is a necessary next step for long-term viability.

The content providers and ASPs have the opposite problem: Their knowledge of upper layer services is quite rich and well developed, but they have little if any network capability. The capital investment required to build a network of their own would be prohibitive, which explains why so many of them are forming alliances or ownership arrangements with network providers. Consider Qwest, for example: as a bandwidth baron, they have a globally deployed optical network, as well as content capability through CyberSolutions, their joint venture with KPMG. Their acquisition of USWest allows them to cover the access, transport, and services waterfront, an enviable combination of capabilities.

There is yet another example of this evolution: the ongoing inversion of the network. As intelligence, capability, and bandwidth move inexorably away from the core toward the edge, the margins of the network cloud expand as they strive to touch the customer. A relatively small collection of centralized functions, delivered by core switches and shared among a large collection of users, tends to treat the customer as a commodity and makes the assumption that their collective service requirements will be largely the same. The process of migrating capability closer to the customer dashes this philosophy on the rocks and enables services to be customized to a highly granular degree. David

Isenberg, author of *The Rise of the Stupid Network*, predicted this evolution years ago in his seminal work. In it, he observes that

> A new network "philosophy and architecture," is replacing the vision of an Intelligent Network. The vision is one in which the public communications network would be engineered for "always-on" use, not intermittence and scarcity. It would be engineered for intelligence at the end-user's device, not in the network. And the network would be engineered simply to "Deliver the Bits, Stupid," not for fancy network routing or "smart" number translation.

These customized services take advantage of the redesigned network and its capabilities and redefine the relationship between customer, provider, and service.

So, what does all this mean? It means that the winners in this game to acquire and keep customers will be those companies that understand the importance of providing highly targeted value based upon a discrete understanding of the drivers behind every customer. Technology, expressed as features, functions, benefits, and so on, is still critically important. However, the people who care about that level of technical detail are not generally the people who make buying decisions or write checks for product purchases. Value takes on many different forms and must be expressed appropriately to each audience. Technology bells and whistles certainly matter—I would never suggest otherwise—but there is much more to the capability equation. The characteristics that represent value to a technician responsible for installing, operating, and maintaining a network are different than those that provide value to a network designer, a beneficiary of the services that the network delivers, or the end user who must make a buying decision about a highly capital-intensive acquisition. Thus, the manner in which each product is represented to the client must be customized to an appropriate degree if the sale is to be successful.

This evolution, illustrated by the separation of duties in the OSI model, the continuum of capabilities of the value equation, and the ever-expanding network and its migration of capability toward the customer, represent the inexorable need for customization, for knowledge about the customer, and the ability to interpret the knowledge and respond with solutions that provide undeniable, targeted value to each and every client's individual needs.

There is an electrical concept called *skin effect* that provides a powerful analogy here. It is a well-known fact that current does not flow in the core of a conductor. Instead, it flows on the surface because of reduced electrical resistance due to the far greater surface area that is available there. This is why stranded wire is a greater conductor than solid core wire: The multiple strands provide enormously greater surface area than a single solid conductor.

The analogy works well here: There is far greater market opportunity surface area at the edge of the cloud than there is within, and although the core marketplace continues to be important, the opportunities, the growth, *the money*, lie at the edge.

The converged telecommunications industry, as we will see shortly, has segmented itself into a hierarchy of functional players that work together to satisfy the changing demands of a remarkably diverse client base. There is a food chain at work through which the various players combine their efforts in a form of value-added service to satisfy customer demands. The collective efforts of the various manufacturing layers and those of the service providers result in the foundation of that which is required to create value in the eyes of the customer. That value derives from stable, reliable hardware and software, functional applications, and a clear focus on the customer's value chain.

A GUIDE TO THE SECOND EDITION

This second edition differs from the first in a variety of ways. Naturally, it is current; I have updated all company information, added new players where appropriate, and deleted those that are no longer germane. I have also gone through the technology section of the book and updated it to ensure currency with regard to how the technologies work, how they are governed by relevant standards, and who the players are that offer them. I have also added a number of new topic areas throughout the book that have emerged as important subjects since the first edition appeared, including regulation, the semiconductor and gaming industries, and the metro networking environment. Some technologies, particularly those that have taken on greater significance, such as optical networking and wireless, have been enhanced.

Finally, the structure of this edition is different from that of the first. Much of the content has not changed. For example, readers of the first edition made it clear that the technology overviews are valuable, so I have elected to keep them (updated, of course). To each of the three convergence sections, however, I have added new material that addresses current industry trends and in some cases marketplace predictions about the various components of the convergence phenomenon. In fact, each major section begins with a discussion of the current, major trends that affect it.

This book has become much more than we originally intended. To all who bought the first edition, thank you for making it so successful. To new readers, welcome to the second of what will be a series of editions that we will continue to publish as long as the material remains relevant.

As before, this edition is linked to a web site where readers can find additional information about convergence and related topics. Please visit **www.ShepardComm.com** and follow the directions on the home page.

As always, I welcome comments and suggestions from readers. Those of you who contributed to the first edition will hopefully see your influence in this book; because of your willingness to share comments with me, the second edition is a better product than the first. Please send comments to *Steve@ShepardComm.com*. Thank you.

Steven Shepard
Madrid, Spain and Williston, Vermont
December 2001

INTRODUCTION

I ask you this: Could there possibly be a more interesting industry to work in than telecommunications? Never a dull moment doesn't come close to describing it. During the past decade, telecomm has taken us on a roller coaster ride that, all claims to the contrary aside, hasn't come close to its grand finale. I have listened with amusement to industry pundits and players declaring that the bubble has burst and that telecomm has reached its proverbial End of Days. Nonsense. Since 1876, when Thomas Watson was shocked out of his shoes by A. G. Bell's voice emanating from the speaker of the Harmonic Telegraph that the two developed, we have continually installed telecommunications infrastructure and developed new applications without pause. In spite of the recent dot-com to dot-bomb conversion (as many have taken to calling it), the supposed glut of installed optical fiber, the failure of company after company, the paucity of new, bandwidth-dependent applications designed to excite and inspire us, and the apparently rudderless movement of the wireless industry, there is still inexorable forward movement and room, therefore, for excitement, growth, and technology-based investment successes.

There is also, however, fear, doubt, hysteria, and inaction, all of which erode confidence and inevitably lead to analysis paralysis on the parts of investors, analysts, and corporate decision makers. This is a complex industry with more operational facets than a marquis-cut diamond, and the task of analyzing each of them and their interdependencies is daunting at best and more often than not a showstopper.

As a professional telecommunications educator and consultant, I make it my business to identify these trends, watch them closely, determine which are significant, and chart any interdependencies that emerge. This introduction is designed to list the most significant of them and provide a vantage point for anyone looking to understand the current direction of the industry at large. The industry includes service providers, device and component manufacturers, regulators, software developers, and of course, the customers that dwell at the top of the technology food chain. Each segment is important and plays a critical role in the ongoing evolution of telecommunications technologies, companies, and services.

Of course, this collection of observations and analyses represents one person's opinion. It is, however, an *informed* opinion, based on interviews with hundreds of professionals in the industry, customers, and the intuition that comes from 20 diverse years in telecomm. As Dennis Miller likes to say, "Of course, this is just my opinion—I could be wrong."

BEGINNINGS: WHAT DO WE KNOW?

Beginning in early 2000, the telecommunications industry has suffered a progressive and colossal meltdown. Several hundred thousand people have lost their jobs; stock options have dissolved and disappeared, and consumer confidence in the telecomm sector as a whole has eroded. JDS Uniphase, once a darling of the investment community, announced the largest loss in investment history ($50.6 billion). Enron, once a massive powerhouse, has declared bankruptcy, the largest in history. Manufacturers as a whole saw a 26 percent decline in second quarter revenues year-over-year, and new orders were down more than 50 percent in the same period. Although capital spending in the United States showed an admirable 25 percent growth rate, average revenues grew only 25 percent, while profits dropped precipitously and return on equity faltered. Between 1996 and 2000, capital expenditures in telecommunications rose from $41 billion to $110 billion—but return on capital fell

50 percent. Finally, since 1984, the domestic consumer price index is up 73 percent, local communications volume is up 71 percent, but long distance prices are *down* 35 percent—not a good sign for the long distance industry.

SO, WHY DID THE CRASH OCCUR?

There are numerous reasons for the so-called telecom meltdown, but some are more glaring than others. Readers familiar with the children's story "The Emperor's New Clothes" will recognize similarities in telecomm's downfall. Telecommunications is an industry that has historically been dominated by its reliance on technology and by a somewhat egotistical and technocentric collection of companies that believed that technology itself was the final product. In the early, heady days of the industry's development, an argument could be made that this was indeed the case—but no longer. As technology became more commonplace and found its way deeper into the public psyche, its presence became an expectation rather than a special thing. There was a time when having a telephone was something to wake up and smile about in the morning—no longer. Today it is a given. Technology is a facilitation mechanism that makes possible a collection of applications that enable customers to position themselves more competitively. It is a business accelerant, a powerful catalyst that helps companies win. Companies don't buy high-bandwidth services for bragging rights. They buy them so that they can reach and respond to customer requests for service faster than the competition. In today's marketplace, time may equate to money, but speed equates to profit. The first company to reach the market wins the game, and the stakes in this game are extremely high.

So, why did the crash occur? Because technologists positioned their technologies as products, and for a while, the market was fooled. When technologists offered multimedia, customers saw television. When they offered DSL, customers heard unlimited capability. When they said wireless Internet, customers were foolish enough to believe them. This contrived anticipation of sexy new services drove demand for bandwidth to an all-time high. Bandwidth barons raced to install fiber in the network core at such a rate that they outstripped the cash available through traditional venture capital sources. Instead, they fell for and were seduced by junk bonds, the same kind that financed the cellular and optical fiber buildouts of the 1990s. In a normal market, there is nothing inherently bad about junk bonds; they are nothing more than a slightly riskier financial instrument that pays a relatively high interest rate, typically issued by a corporation that is so new that it does not have an established earnings history.

However, this was not a normal market. There was a mania associated with it; any company that purported to have the next big thing became an instant money magnet. Venture capital flowed like a class 5 river because venture capitalists were blinded by the promises of technology. Meanwhile, the market clamored for services, applications, and solutions, and few listened.

Soon, a gap began to develop. Like the little boy who was naïve enough to proclaim the king naked, customers began to question the growing disparity between whizbang technologies like 3G, DSL, the wireless Web and cable modems, and the services that they were not yet able to deliver. It soon became clear that the gap was not going to narrow anytime soon, and customers' academic questions became serious financial concerns. Complicating the situation was a set of Byzantine regulatory decisions, signed into law in 1934 and slightly modified in 1996 in the United States, that, although promising to open the local telecommunications marketplace and introduce unfettered competition, in fact accomplished the opposite. The phenomenon was not limited to the United States. Confidence fell; investors pulled their investments. Debtors, left with huge loans, found themselves having to pay off accumulating debt with appreciating dollars, while cash flows declined and collateral evaporated. Major manufacturers like Lucent and Nortel, long accustomed to serving as financing entities for startups, suddenly found themselves holding the paper on enormous quantities of debt that would never be repaid. Banks turned off the capital spigot, refusing to roll loans over because of default jitters. The industry ground to a halt, and the sun set on a lot of rising stars.

EVOLUTION

Since the divestiture of AT&T in 1984, the telecommunications services industry has metamorphosed from a small, easy-to-understand collection of companies to a rather complex jumble of players, all operating with varying degrees of success. On December 31, 1983, there were 22 *Bell operating companies* (BOCs), two long distance providers, one significant hardware manufacturer, and about 1,500 independent telephone companies that provided telephony service throughout the United States. On January 1, 1984, the 22 BOCs coalesced into 7 *Regional Bell Operating Companies* (RBOCs); the other players remained the same for quite a long time. In 1986, Sprint, an outgrowth of United Telephone, added its name to the list of nationwide long distance providers, and over time the cadre of hardware manufacturers grew as awareness of customer demands for choice grew. The same evolution took place at the local level. Initially competing in major metropolitan markets with the seven RBOCs were *Teleport Communications Group* (TCG) and *Metropolitan Fiber Systems* (MFS), the first pejoratively-named bypassers, soon relabeled alternative access providers. These companies built proprietarily owned optical ring networks in the central business districts of large cities and competed effectively with the incumbent providers by offering redundant dual entry and an all-optical transport infrastructure. They flourished because they had the luxury to cherry-pick their markets. They did not labor to develop a presence in Dime Box, Texas, or Scratch Ankle, Alabama; they did, however, establish beachheads in Dallas and New York City, Minneapolis-St. Paul and San Francisco, Seattle and Miami because, as bank robber Willie Sutton was wont to observe, "That's where the money is."

Over time, these local loop competitors became known as *competitive local exchange carriers* (CLECs), while the RBOCs became known as *incumbent local exchange carriers* (ILECs). The line, however, blurred somewhat: Some ILECs announced their intentions to enter the territories of other ILECs, effectively becoming de facto CLECs in their own right. Over time, the ranks of the CLECs swelled to several hundred in the United States alone. Meanwhile, other segments emerged from the service shadows. Beginning with the arrival of the Internet and Web in 1993, *Internet service providers* (ISPs) grew like mushrooms on a summer lawn—more than 5,000 in the Unites States alone. Independent wireless providers like Winstar and Teligent sprouted in cities. A new segment, called *data local exchange carriers* (DLECs), also emerged with names like Northpoint, Rhythms NetConnections, and Covad. Taking advantage of a regulatory decision intended to further increase competition at the local loop level, they offered DSL service over existing local loops. In other words, a customer could purchase their phone service from the local ILEC, but had the option to buy their broadband access from someone else. Pure bandwidth providers (sometimes called *bandwidth barons*) like Qwest, Global Crossing, and 360Networks, riding on massive installed bases of long-haul optical fiber, offered ridiculously low prices on transport. Cable companies, having largely completed the digital and optical conversion of their local distribution networks, began to offer two-way interactive services, including Internet access. Hardware manufacturers like Cisco, Nortel, Alcatel, and Lucent, feeling the growing pain of competition, built advertising campaigns around the technology that underlay their products.

Meanwhile, long distance companies began to feel the pinch of competition from the so-called bandwidth barons. Soon it dawned on them that their vaunted service was in fact a commodity, and that like any commodity, the only way to be the preferred vendor was to have the lowest price—which they simply could not do when confronted with the massive over-installed base of the bandwidth barons—particularly when faced with the regulatory structure under which they were required to operate. As we will see, that same regulatory structure resulted in problems for other segments as well.

Other cracks in the telecommunications armor began to appear as time passed. The companies that made pure technology plays, such as the ISPs, the standalone wireless companies, and the DLECs, began to realize that their offered service was in fact not a service at all—it was technology and commodity technology at that. Visions of world-changing capabilities and applications, riding on the promises of technology, began to show their limitations, and customers and investors began to ask hard questions.

AFTERMATH

Today, the industry make-up is dramatically different. For the most part, the CLECs and DLECs are a dying breed. The biggest names—Covad, Northpoint, and Rhythms—are operating under bankruptcy protection or gone. The herd of pure ISPs, such as PSINet, has been dramatically culled. Teligent, Winstar, and Metricom, the pure wireless technology providers, have failed. Hardware manufacturers are shedding employees and restructuring, the universal panacea, and are awash in inventory. AT&T, WorldCom, and Sprint, the nation's premier long distance providers, are questioning their role in the evolving industry, and their nemeses, the bandwidth barons, are faltering.

Three segments of the industry are glaringly absent in the preceding paragraphs. During this technological bloodletting, the ILECs and the mixed-breed content and service providers have escaped with minor cuts and bruises. The ILECs/*post, telephone, and telegraph* (PTTs)—Bellsouth, Verizon, SBC, USWest/Qwest—are wallowing in cash and planning their capital expense budgets for the coming year. AOL-Time Warner, the biggest of the online service and content providers, grows apace and enjoys a constantly reinvigorated role in the marketplace, and Microsoft, with its .Net initiative, positions itself for the biggest gamble—and perhaps upside—in its 23-year history.

THE PIVOT POINT: CUSTOMER VALUE AND HOW TO MEASURE IT

The painful realization that technology is not the answer to the market's challenges is reverberating throughout the telecommunications industry. Technology is a facilitation mechanism, nothing more. It is powerful and necessary, but ideally should be invisible in the eyes of the end customer. The customer should see and be impressed by the effect of well-used and optimally positioned technology, not by the technology itself.

The brochures distributed by manufacturers at trade shows and via the Web typically describe their offered products in terms of three characteristics: features, functions, and benefits. Such characteristics as heat dissipation, footprint, power consumption, mean time between failure, accessibility, and the ability to upgrade easily are all described here. Unfortunately, these parameters mean nothing to senior managers in an organization that must make capital expense decisions for their companies. Their focus is on such factors as competitive positioning, the ability to resolve customer business challenges, and the resulting ability (or not) to preserve shareholder value. They just don't care about the amount of heat a box produces or the number of square feet the box requires in the central office. This is not to imply that these factors are unimportant: They clearly are, but they are important to office engineers, installation technicians, and maintenance personnel, and by and large these employees do not sign checks. They may influence the decision, but they do not have the final say. It is critical, therefore, that manufacturers who want to be noticed by the people who *do* sign the checks position their products using messages that carry information those people care about.

SO, WHAT'S THE ANSWER?

The efforts by service providers, content providers, manufacturers, and software developers to satisfy customer demands are facilitated by an ongoing phenomenon that they would do well to pay attention to. For over two years now, convergence has held sway as an inexorable force that is guiding the development of the technology and communications sectors. Those companies that choose to heed its warnings and caveats are succeeding in their efforts; those that do not are disappearing.

Telecommunications convergence is a three-part phenomenon. Technology convergence represents the unstoppable drift toward a packet-based infrastructure with particular attention being paid to the *Internet protocol* (IP). By creating a converged network infrastructure, service providers can enjoy the benefits of lower overhead costs associated with network operations and the ability to offer unified services across a common, low-cost, simplified platform. Technology convergence, then, is a facilitation mechanism.

Company convergence grew out of the realization that as customer demands grew and became more diverse, offered services needed to evolve as well. Service providers and manufacturers were thus presented with a choice: They could create the necessary enhanced capabilities in-house, or they could go out and form an operational relationship with someone who already had the capability, either through an alliance, partnership, or outright acquisition. Company convergence, then, is a convenience mechanism.

Services convergence represents the ultimate goal that all players strive for: the ability to offer the customer exactly what they are seeking in the way of services and capabilities, ideally via a converged network and technology infrastructure, facilitated by carefully planned company convergence efforts.

Companies that heed the words of the convergence oracle will benefit from its wisdom. Customers no longer care about the inner workings of networks or the devices that live within them; they care about what those devices and agglomerations of technology can do for them in terms of enhanced competitive advantage, customer base preservation, brand enhancement, and market longevity. Service providers and manufacturers that make it their mission to choose technology as their competitive advantage will most assuredly be relegated to the dim corners of market disinterest. This is not melodramatic; it is measurably true.

There is a vast sea change underway within the technology and communications sector, made up of many different, but often highly interrelated forces. The purpose of this book is to examine these forces and study their interrelatedness so that some sense can be made of them and the impact they have on the greater industry's overall direction. Coincidentally, all of them fit, one way or another, into one of the three convergence segments. Furthermore, each segment sports a collection of key observations that characterize it; they are described in the following list and will be discussed in greater detail in each section of the book.

Technology Convergence

- The Internet's influence is far from over.
- Broadband access will continue to be important.
- The software industry will take on a more critical role as the players evolve.
- Optical technology will be the next big techno-hero.
- The semiconductor industry's influence will grow substantially, and products will become more application specific.
- In spite of the hype, 3G doesn't cut it. Wireless, however, is a critical component of broadband success.
- Mobile appliances will become a significant marketplace.
- The functional migration from the edge of the network to the core is underway.
- The metro environment will become a central marketplace in the next three to five years.
- Network management will take on an increasingly important role.

COMPANY CONVERGENCE

- Companies will continue to form new and often strange alliances, and the number of players in the herd will continue to shrink.
- A new regulatory environment is needed and will be created.

SERVICE CONVERGENCE

- Content was, is, and will be king. Period.
- *Storage area networks* (SANs) will become centrally important as the core-to-edge evolution continues.
- The electronic gaming industry will serve as a major technological bellwether.

In the pages that follow we will examine each of these components of the convergence troika, the drivers within them, and the manner in which they are interrelated.

PART ONE

THE TECHNOLOGIES

Any sufficiently advanced technology is indistinguishable from magic.

—*Arthur C. Clarke*

INTRODUCTION

As we promised in the introduction of the book, each of the three major sections of the book will begin with a collection of predictions, prognostications, and observations about the most important issues affecting that particular piece of the convergence triad. These follow the Dennis Miller rule: "These are only my opinion (informed, of course)—I could be wrong."

THE INTERNET'S INFLUENCE IS FAR FROM OVER

Internet Protocol (IP) transport, specifically the Internet, represents the architecture of the future multimedia network. As we will see a bit later, the conversion from a largely circuit-switched infrastructure to a packet fabric will begin in the network's core and migrate outward to the edge. Packet switching is the protocol of the new network paradigm; IP, *Asynchronous Transfer Mode* (ATM), and *Multiprotocol Label Switching* (MPLS) will collaborate to ensure that multiservice networks will be able to offer variable *quality of service* (QoS) levels as customers demand variable pricing structures for offered services. Furthermore, IP will continue to enter previously impenetrable areas such as voice as QoS becomes real. Although some may claim that the Internet's influence has reached a plateau, others differ; Larry Roberts, one of the true Internet pioneers who played a pivotal role in the early development of protocols used in the online world, recently observed that in the past year, Internet traffic has quadrupled. Eighty percent of that traffic is corporate; the so-called dot-coms contribute less than 5 percent. Traffic from game playing and chats will become large contributors to the overall traffic makeup, provided broadband access becomes available.

Another area of great interest—and perhaps influence—is e-commerce. As a general rule, intelligence in online commerce will rule as shopping bots take over the onerous tasks of product search and price/feature comparison, purchasing, and delivery logistics, all consumer (not provider) driven. Money will fundamentally change as digital cash becomes widely accepted and trusted thanks to evolving (and effective) privacy and security overlays as well as the acceptance of trusted third-party players. Additionally, taxation practices will have to evolve and the changes will be profound, given the global nature of the online marketplace. This marketplace is real; Nicholas Negroponte, director of MIT's Media Lab and author of *Being Digital*, was right. The complete digitization of certain types of products is underway, converting them from physical entities to logical entities for delivery to the customer. Music, video, software, and games are shedding their dependency on physical media for delivery to the customer and are relying instead on the growing availability of broadband access and the customer's ability to download the product within a reasonably short time interval. *Make no mistake*: success is fundamentally dependent on the widespread availability of broadband access.

Another fact of digital life is that *Internet service providers* (ISPs) will become irrelevant because pure technology plays do not succeed in this marketplace. *Commercial service providers* (CSPs) and *business service providers* (BSPs) will become centrally important because they combine access with some form of desirable online content. Furthermore, end-to-end transaction management will become an offered solution, and all players will vie for a piece of this market because it is close to the customer and therefore lucrative.

Protocols continue to play a key role. HTML has long been the preferred formatting language for Web-based activities, but this will soon change. *eXtensible Markup Language* (XML) and *eXtensible Style Language* (XSL) will replace it because they have the capability to separate the content from the delivery mechanism. The initial players in this evolution will be banks, access and transport companies, and software/hardware vendors. Companies like *America Online* (AOL) and Microsoft will prove to be pivotally important as the vanguard of change.

The Internet deserves a great deal of attention because it has served for quite a few years as a catalyst for new ideas and thought. As a business tool, however, it has some limitations. The functional decentralization of the Internet yields a lack of reliability and responsibility that can be, at best, vexing; however, it does promote a high degree of innovative behavior. As applications emerge to take advantage of the newly intelligent Internet, application-specific access and transport protocols emerge as an opportunity for innovation. This is what Microsoft's .Net initiative is all about: become the monopoly network and application provider, recognizing that with the proper packaging, the two become functionally indistinguishable. Microsoft's initiative relies on network-based subscriptions for access to content, access device agnosticism, a PC-centric model, and a reliance on XML. The model should be watched closely.

In computers, the resource to be wasted is transistors because this practice enables *everyone* to have a low-cost computer. In networks, the resource to be wasted is bandwidth because this enables everyone to have a dedicated, always-on connection. Broadband is a key success measure as evidenced by the sustained growth in *Digital Subscriber Line* (DSL), cable, and wireless local loop access options that continue apace as customer demand for high-speed access increases in response to growth in multimedia applications. One clear sign of the importance of this area is that semiconductor manufacturers have focused on this segment as a major piece of the market. Furthermore, the functional development of the network continues to move inexorably toward the need for high-speed access to complement the network's high-speed core. This paucity of service is real; only 10 million U.S. subscribers have broadband access today.

The three technologies with the greatest chance of bringing broadband to the masses are DSL, cable, and wireless. All, however, suffer from daunting challenges. Cable suffers from the old adage that observes, "In its success lie the seeds of its own destruction." Cable wrestled for years to overcome the market perception of inadequacy as a delivery medium for interactive voice and data services. Now, as Internet use has grown and cable modems have come into their heyday, another challenges has arisen. The interactive cable network relies on a shared bandwidth infrastructure, which means that the more people use it, the less bandwidth is available for each subscriber. DSL, meanwhile, has its own challenges. In North America, more than half of all local loops are deployed over digital loop carrier systems that make available a maximum of 64 Kbps of bandwidth for each subscriber. DSL, therefore, cannot be deployed over this infrastructure without significant technological change. Additionally, the technology is hobbled by distance limitations and highly publicized installation failures.

Finally, wireless is positioned as the ultimate winner in the quest for broadband customers, but only if adequate spectrum is made available by government agencies and applications evolve to drive demand. In the United States, the FCC is currently considering the addition of spectrum in the 1910- to 2400-MHz range. This spectrum would have to be taken away from satellites, the *Multichannel Multipoint Distribution Service* (MMDS), unlicensed *personal computing services* (PCS), and amateur radio, and would be made available immediately to companies ready to deploy *third generation* (3G) infrastructure and services. There is, however, a further caveat. According to the Gartner Group, in order for 3G to succeed, 50 percent of the population must have access to 75 percent of the offered services, 5 percent of the population must have the most recent devices, and an obvious and widely lauded killer application must be evident. Reality rears its ugly head, however; today, only 18 percent of U.S. and Canadian subscribers have access to broadband, while 20 percent simply cannot get it. This lack of broadband access cannot be enabled to continue; studies have shown that widespread broadband local loop technology can add three billion work hours annually to the North American economy, a number that is well worth considering seriously.

Driving the demand for bandwidth is content, which continues to become more and more media-rich, expanding demand for high-bandwidth access. Couple this with growth in the number of home-based workers, *small-office/home-office* (SoHo) workers, and telecommuters, and it is easy to see why the demand for bandwidth is expanding.

Closely linked to the content market is the home/office access arena. Premises technologies, such as the various flavors of 802.11, HomePNA, *wireless applications protocol* (WAP), and Bluetooth, will contribute to the success of broadband access, inasmuch as they will expand the variety of content that is reachable.

THE SOFTWARE INDUSTRY WILL TAKE ON A MORE CRITICAL ROLE AS THE PLAYERS EVOLVE

Software has never been particularly visible when standing in the shadows of such technological luminaries as optical transmission hardware and blazingly fast computers, switches, and routers. That, however, is changing. As Larry Ellison's vision of the network becoming the host for software applications approaches reality (something Microsoft has taken an enormous gamble with in its .Net initiative), applications-support software will become centrally important. It will include *business-to-business* (B2B), *business-to-consumer* (B2C), *peer-to-peer* (P2P), *application service provider* (ASP), security packages, and more. It will be the element that causes hardware to perform at its highest level of capability and offer support for a diversity of service types.

OPTICAL TECHNOLOGY WILL BE THE NEXT BIG TECHNO-HERO—AGAIN

The growing demand for bandwidth has driven service providers to invest deeply in optical infrastructure in the core of the network and to a lesser degree in metro environments. This creates a layered have-and-have-not problem. The core is now over-provisioned, as evidenced by the focus on the glut of dark fiber that now exists. Unfortunately, the edge is massively *under*-provisioned, resulting in a very uncomfortable transport disparity. Optical switching, routing, and multiplexing, as well as unimaginably fast transports, are realities today; unfortunately, they encounter a wall when they try to leave the core and head for the customer over the largely narrowband local loop. Ultimately, this problem will be resolved, but until that time, optical technology remains a transport technology rather than an access solution.

THE FUNCTIONAL MIGRATION FROM THE CORE OF THE NETWORK TO THE EDGE IS UNDERWAY

There was a time when universally centralized functions made sense for economic and technological reasons—no longer. High-speed networking has become a cost-effective reality, and service providers have realized that the closer the delivery point of a service is to the customer, the better the service can meet the demands of the customer. As a result, intelligence is migrating from the core to the edge, and intelligent, high-speed devices have become centrally important components in the overall network. The migration of intelligence from the core to the edge is necessary. IP, MPLS, *Media Gateway Control Protocol* (MGCP), and *Signaling System Seven* (SS7) suddenly become peer protocols as they struggle with the demand for QoS-aware capabilities.

Furthermore, the migration of network functionality to the edge is also a requirement. By placing network resources and capabilities at the edge of the network, they are closer to the customer and can, therefore, be customized to meet individual client requirements.

It is interesting to note that network convergence requires network *divergence*. Examples are becoming more and more common: the migration of the *digital subscriber line access multiplexer* (DSLAM) from the central office to the neighborhood or business park, the migration of intelligence into customer-resident *Private Branch Exchanges* (PBXs), the deployment of remote switch modules, and distributed, edge-based signaling.

IN SPITE OF THE HYPE, 3G WIRELESS DOESN'T CUT IT

3G is a technology that is poorly positioned—a technology in search of a problem to solve. Originally advertised as the facilitator of the wireless Internet, customers have come to realize that it is nothing of the sort. A recent study showed that the *average* number of clicks and scrolls a subscriber had to issue to pull up a stock quote on a Web-enabled cell phone was 22—down from the 26 required on the previous year's model. The combination is doomed to fail inasmuch as it represents a deadly combination of immature, not-yet-ready-for-the-market technology, poorly established expectations, slow, spotty connections, useless screens, and non-existent content. Comparing the experiences of surfing the Web via a computer and surfing it with a cell phone is a laughable exercise—there *is* no comparison.

Key to the success of broadband, obviously, is spectrum availability. The spectrum battles that are underway between broadcasters and telecomm players create enormous problems that cannot be overcome without a shift in regulatory policy. Television has always been a sacred cow in the United States and Canada, and it has always enjoyed the rights of first refusal for newly available spectrum. As a result, they have amassed huge bunkers of bandwidth. The fact that they only use a relatively small percentage of what they hold appears to be irrelevant. It's high time federal regulators recognize that sacred cows sometimes make very good burgers and redistribute spectrum as appropriate to level the services playing field.

Of course, exceptions abound. The enormous fees paid by service providers for 3G licenses have turned out to be proverbial albatrosses around their necks. Some of them, such as Sonera, are attempting to give the licenses back to the governments that sold them in the first place. Once again, 3G appears to be a technology in search of a problem to solve, a problem in the form of the ever-elusive killer application.

MOBILE APPLIANCES WILL BECOME A SIGNIFICANT MARKETPLACE

Handheld functionality represents a significant share of the application and device market. It is one of the few segments that is truly market-agnostic, largely because of the plethora of applications available for the devices. Most users have replaced the paper functionality of the three-ring planner, while many have gone further, adding e-books, games, digital camera modules, and wireless functionality. What will emerge are devices that are actually useful, built around the functionality that the customer wants to use, not what the technologists want to deliver. Configurable *personal digital assistants* (PDAs) will rule.

Adding to this successful evolution will be wireless protocols such as Bluetooth, WAP, and especially 802.11. They will facilitate the entry, uptake, and success of mobile appliance-based applications, provided they have access to adequate bandwidth and can support the suite of applications that customers actually want.

THE COMPUTER AND PERIPHERAL DEVICE INDUSTRY SEGMENT WILL EVOLVE IN STRANGE AND WONDERFUL WAYS

With the arrival of such technologies as InfiniBand, the ability to create a *global area network* (GAN) will become a reality—

and a much-vaunted capability. This will redefine network computing as we know it today as the physical locations of processors and peripherals, particularly storage devices, become immaterial. Similarly, the storage area network concept will take advantage of this evolution and will become central to the success of businesses. As Frances Cairncross observed in *The Death of Distance*, physical distance and geography become immaterial.

THE METRO ENVIRONMENT WILL BECOME A POWERFUL MARKETPLACE IN THE NEXT THREE TO FIVE YEARS

It is almost unthinkable that only 10 to 12 percent of all business facilities in the United States and Canada are wired with fiber today. Once again, broadband access, this time in the business environment, is hobbled by the lack of a simple infrastructure component.

There was a time when a corporate headquarters was exactly that—a single, monolithic building that housed all of the corporation's employees. Today that environment is evolving such that multiple corporate locations within a metro area are the norm as companies learn to place staff close to major customer clusters. Unfortunately, this shift in real estate philosophy brings with it an additional challenge, this one technological: They need to be interconnected. Thus, metropolitan area networking will experience serious growth and interest in the next few years. Companies like Yipes and Telseon will rule this marketplace with such offerings as native Ethernet transport. Equally important is the fact that the overall connectivity architecture for metro will be mesh-based; rings will be used primarily in long-haul networks.

THE SEMICONDUCTOR INDUSTRY'S INFLUENCE WILL GROW SUBSTANTIALLY AND PRODUCTS WILL BECOME MORE APPLICATION SPECIFIC

The introduction of the microprocessor offered one key advantage: It brought the flexibility of computer programming to non-computer applications, resulting in the widespread proliferation of computing capabilities. At the same time, it made embedded, silicon-based software trivial.

The next stage was the programmable logic device, such as the *Field Programmable Gate Array* (FPGA). This enabled designers to use the most abundant resource available (transistors) to reduce the requirement for the scarcest resource available (system designer time). The goal of this overall effort was to convert processors, peripherals, and other devices to custom design files known as *intellectual property* (IP).

Moore's Law, based on the mandate to put more transistors on less silicon real estate for a fraction of the cost, permitted the migration of such peripheral devices as the *Moving Pictures Expert Group* (MPEG) encoders/decoders, serial channel units, amplifiers, and multiplexers from external positions to the chip itself, resulting in exponential reductions in space utilization (real estate). This resulted in an expansion of microprocessor-controlled devices, the functional result of which is the *application-specific integrated circuit* (ASIC).

The result was a move away from wholesale chipset manufacturing to more customer-specific chipset designs. Thus, *System-on-a-Chip* (SoC) functional models will begin to appear in greater numbers and with far greater specificity—but with far greater financial rewards as well. Intellectual property (the other IP) will become the *real* moneymaker in the industry and will become coveted and carefully protected.

The semiconductor industry will most likely be the first segment to recover economically because it represents the plankton at the bottom of the food chain upon which all higher layers rely.

NETWORK MANAGEMENT (NM) WILL TAKE ON AN INCREASINGLY IMPORTANT ROLE

NM, as opposed to element management, has already become a focal point and rallying cry within the industry, more because of the number of companies that do not have it than those that do. Many companies in the industry offer element managers with their products, but these are device specific and often based on proprietary software infrastructures. Furthermore, because of their proprietary nature, they do not readily integrate into higher-level management systems designed to surveil the entirety of the deployed network. One thing is certain: The company that releases a single-seat, vendor-agnostic network manager that is truly capable of managing an entire network will immediately rise to the top of the credibility chart and enjoy substantial returns on their own hardware offerings. Because of vendor specificity and a commonly seen lack of understanding the difference between element and network managers, the critical process of managing a deployed infrastructure is far more onerous, costly, and complex than it needs to be. This area needs the attention of the industry.

We now turn our attention to the technologies themselves what they are, how they work, who supplies them, who uses them, and why.

CONVERGENCE: THE TECHNOLOGIES

As the foundation layer of the convergence troika, telecommunications technologies can be subdivided into two interrelated segments: *access*, which includes not only the actual circuit used to connect user devices to the network, but the access device technologies themselves, and *transport*, which describes the fabric of the *wide area network* (WAN). Two other technology areas, signaling (the protocols and procedures used to establish, maintain, and terminate voice, data, and video transmissions across the network) and NM (the set of procedures used to operate, maintain, and provision network resources) will be discussed in the Services Convergence section later in the book.

We will approach our examination of communications technology from the point of view of each of these because in reality, all have some bearing on the capability of the network to provide services to the customer. *Access*, including the device(s) used by the customer to gain access to the network itself, addresses local loop technology options that a customer might employ to enter the network cloud, such as traditional twisted pair, *Integrated Services Digital Network* (ISDN), DSL, cable modems, and wireless services. This region of the network is particularly important because it touches the customer; whoever controls access may well control the customer, as evidenced by the fierce battle that is underway between incumbent local service providers, long-distance carriers fighting to reenter the local services market, and their emerging competitors, the multiservice cable providers.

Transport defines the set of technologies that make up the fabric of the network cloud, including circuit switching, traditional packets, frame relay, ATM, and IP. As the marketplace evolves to demand more in the way of wide-area, end-to-end services (particularly broadband data), the nature of the transport fabric becomes rather critical, particularly in converged networks where it must capably transport multiple traffic types that require varying degrees of service quality. Furthermore, it is evident that a pure, single-technology network is probably not likely, at least not in the near term. Most likely, the network of the foreseeable future will comprise of a hybrid of multiple technologies capable of supporting the diverse transport- and service-quality requirements of an increasingly complex set of customer applications. Ideally, of course, customers will remain blissfully unaware of this technological cornucopia and will have the luxury of focusing on their own customer requirements rather than on the functionality of the technology that underlies their application set.

Signaling is one of the most important yet least understood aspects of telecommunications networking. In order to establish, maintain, and tear down a connection, some form of signaling protocol must be employed. Not only must the signaling system accomplish the three functions just mentioned, it must also be able to dip into distributed databases to retrieve subscriber information and establish the connection according to the information it finds there. For example, SS7 employs customer databases known as *service control points* (SCPs) that store calling feature information, credit card data, 800 number translations, and a plethora of other data points that can be used by the network to quickly and accurately establish a connection for the subscriber while invoking a preferred set of features. Catalog companies routinely rely on this technology to give themselves the ability to greet a caller by name and discuss that caller's recent purchase history. It gives fast food delivery companies the ability to track customer purchase histories and target those who ordered once and never called again. They can identify those callers and send them a special promotion to try to win them back. Signaling is the technology that makes this possible. Today it is enjoying a greater degree of scrutiny because of the need for signaling systems to provide their capabilities to both traditional *public switched telephone network* (PSTN) environments and emerging IP-based wide area transport networks.

In the multimedia environment, signaling becomes centrally important because it enables a multisession call to be established. For example, a voice conversation between two people may reach a point where the callers need to invoke a videoconference session. Through multimedia signaling, they can easily do so without affecting the ongoing voice call. Furthermore, they can add other sessions, such as fax or data as required, simply by requesting that the signaling network identify the necessary bandwidth, reserve it, qualify its capability to provide acceptable QoS, and establish the path between the identified endpoints. In the evolving call center environment, this capability to establish multimedia sessions is becoming an important part of the customer service equation. Today this is most commonly accomplished in the wide area environment with ATM, a network technology (discussed later) that provides QoS control through an embedded set of service definitions that can be manipulated and adjusted as required.

NM, often referred to in telephone company parlance as *Operations, Administration, Maintenance and Provisioning* (OAM&P), provides the toolset required for network hardware and software to deliver what the customer demands in the way of network services. *Operations* is an umbrella term that includes the other three: *Administration* denotes internal processes required to keep the network functioning properly; *maintenance*, the task of monitoring the network, detecting faults when they occur, and repairing them; and *provisioning*, the process of designing and building network infrastructure throughout the operations area so that when customers request service, the network is in place to deliver it in a timely manner. One word of warning: All manufacturers include as part of the delivered product an element management module that enables service providers to remotely monitor, configure, and troubleshoot network elements as required once they are installed in the network itself. Many manufacturers and service providers alike fall victim to the trap of referring to element management as NM, when in fact the two are very different. Element management is exactly that—the ability to manipulate discrete elements of the network as required. NM, on the other hand, refers to the ability to monitor and manipulate the actual services being delivered by the network as a whole. NM is a powerful differentiator that demonstrates a vendor's cognizance of service as the ultimate product. Element management is far more primitive. Service providers care about the ability to manage network elements; customers do not.

All four of these functions will be examined in detail later in the chapter.

ACCESS TECHNOLOGIES

Historically, access has referred to that subset of the network that connects the customer to the wide area transport fabric, whether it be the PSTN, an ATM or frame relay network, the

Internet, a private IP installation, or some other solution. In its earliest incarnation, access referred to the local loop, the twisted copper pair that has historically provided connectivity between the customer's access device (telephone or modem) and the network cloud.

When the telephone network was originally conceived in the late nineteenth century, it was designed to transport the limited bandwidth requirements of the only traffic type in existence at the time—voice. By the middle of the 1920s, thanks to the work done by such Bell Labs luminaries as Harry Nyquist, it became known that the human voice comprises a rich spectrum of frequencies that range from 20 Hz to approximately 20 KHz. It was also known that although this broad spectral range contributed to the extreme richness and timbre of human speech (think James Earl Jones), it was not required to recognize or understand the person on the other end of the connection. This was actually a good thing. Had network engineers determined that voice required the provisioning of 20 KHz of bandwidth to every telephone call, the cost to build a network capable of doing that would be enormous and highly impractical. As a consequence, engineers determined that as long as each telephony conversation was given 4 KHz—a far cry from 20,000—the person at the other end of the line could be both recognized and understood. For the delivery of voice, this proved to be adequate.

Time went on, however, and the legacy analog network evolved. By the late 1950s, work was being done on digital transmission, and in 1962, the first T-Carrier system was installed to carry digitized voice. This system, operating at 1.544 Mbps, carried 24 simultaneous conversations over two pairs of wire, dramatically reducing the plant requirements typically needed for that many subscribers. This was a great step forward because it enabled network engineers to take advantage of the magic of digital transmission. Specifically, it meant that transmission lines could be made cleaner and that pair gain was achievable, that is, the capability to transmit multiple conversations across a single facility.

Let us dispel a common misconception here. It is often said that digital facilities experience fewer errors than their analog counterparts. In fact, this is not true. Both forms of transmission suffer from the same impairments—impulse noise, crosstalk, thermal noise, harmonic distortion, and a variety of others. The difference lies in how they deal with those impairments. In order to understand that difference, we need to first conduct a short and painless discussion of the difference between signaling and content, that is, the difference between the way in which we represent the information on the transmission line and the information itself.

In common network parlance, the word analog means continuous. An analog wave is a wave that is continuously varying in time. Here's an example: Imagine if you were to use an analog light meter—that would be the one with a needle—to create a graph of the intensity of sunlight from sunrise to sunset. The graph would be continuously variable over the 12-hour period and would look like the drawing shown in Figure 1-1.

Now do the same thing, but this time use a digital light meter—the kind with digital readouts rather than a needle. As you graph the intensity curve, it will not be continuously variable over time, but will in fact comprise a series of discrete

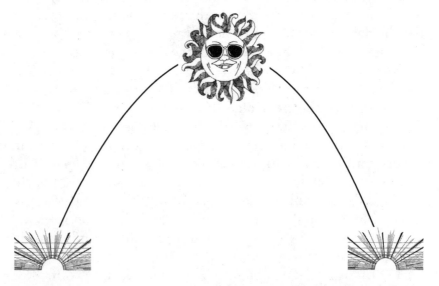

FIGURE 1-1   Analog graph of sunlight intensity in a single day.

stairsteps, as shown in Figure 1-2, that represent the measured intensity of the sunlight.

The analog wave represents an infinite series of data points that convey the required information, while the digital (the word, by the way, means *discrete*) wave comprises a more limited number of data points to convey the same information. Obviously, the digital wave represents an approximation of the actual intensity of the sunlight at any point in time, but it should be obvious to the reader that if enough data points are captured, the digital wave will do an adequate job of recreating the information.

The techniques described here represent *signaling techniques*. Analog signaling uses a continuously varying waveform to represent the information transported, while digital transmission uses a discrete collection of data points to do the same thing. Both are effective, and although it could be argued that the digital technique is really only an approximation of the actual content, we know from further work done by Nyquist in the late 1920s that there is an optimal number of data points

FIGURE 1-2   Digital graph of sunlight intensity in a single day.

that must be generated to create a faithful representation of the original signal.[1]

Analog waves can be modified in a variety of ways to cause them to carry information. Specifically, there are three characteristics of the wave that can be modified: the amplitude, the frequency, and the phase. The amplitude is a measure of the loudness of the signal—it can be high or low. The frequency, similarly, can be high or low, and the phase, a somewhat more complex concept, can be modified to represent the location of the wave form relative to a reference point at any moment in time. These are illustrated in Figures 1-3a-c.

It is important for the purposes of this discussion to remember that we are still talking about how we represent information on a transmission facility, not the nature of the information itself. In the original analog telephone network, analog waves (the actual voice impinging on the microphone of the telephone) caused a current to vary in analog fashion, which in turn passed through the fabric of the network to the telephone on the other end. At the receiving phone, the same varying

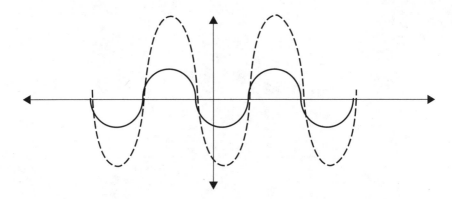

FIGURE 1-3A   Amplitude modulation.

---

[1]Specifically, Nyquist postulated that as long as the original signal is digitally sampled at a rate that is twice the rate of the highest frequency in the original signal, the resulting series of discrete samples will be adequate to represent the original signal. For example, the upper end of the so-called voice band is 4 KHz; according to Nyquist, then, as long as the signal is sampled 8,000 times per second, the resulting digital signal will accurately and faithfully represent the analog signal. This observation is known as *Nyquist's theorem*.

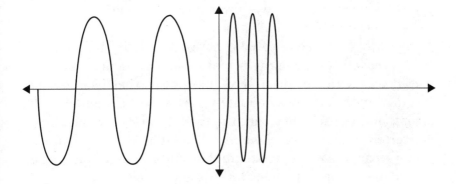

FIGURE 1-3B   Frequency modulation.

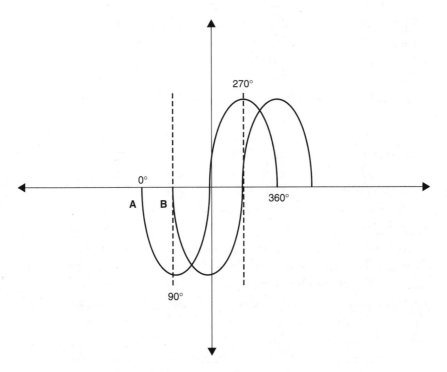

FIGURE 1-3C   Phase modulation.

current caused the speaker in the handset to vibrate analo-
gously, creating sound that the (human) receiver was able to
hear. The point is that the information going into the network

was analog in nature; it was transmitted in analog format, and it was delivered as an analog wave.

Now let's replace the analog network with a digital network. In this case, the incoming signal (voice) is analog; it is converted to a series of discrete digital pulses that (according to Nyquist) faithfully and adequately represent the original analog signal, and at the other end, the digital pulses are once again converted to an analog signal for delivery to the receiving customer's ear. Please understand: The information being transported is analog. By choosing to digitize it for transmission, however, we represent it in a digital fashion. The information itself is still analog and always will be.

So why did we choose to go through this technological nosebleed? Because digital and analog technologies currently coexist peacefully and will for some time to come. Just because information is transported over a digital network does not mean that the information itself is digital. Consider, for example, the *Asymmetric Digital Subscriber Line* (ADSL), which provides high-bandwidth transport across a standard two-wire local loop. With ADSL, in spite of the fact that the information being transported is oftentimes digital, the actual transmission facility is analog. The devices at either end of the circuit are in fact modems, not digital transmission devices.

So back to our original point about the relative goodness of analog and digital facilities. Digital facilities are not cleaner than their analog predecessors—in that regard, there is no difference. The difference and the relative goodness of the two lies in how they deal with the impairments that affect the data they transport. In both systems, signals have a tendency to degrade or weaken over distance. They also have a tendency to be affected by spurious noise signals, and those noise signals are, unfortunately, cumulative. The farther the signal is transmitted, the more noise it accumulates.

In analog systems where there is a continuous, infinite stream of useful data points, errors can often become indistinguishable from the data itself. A 30-ms impulse noise hit, for example, can look like a data point and therefore result in an error in the information stream. Furthermore, as the actual signal weakens over its transmitted distance, the level of the desired signal weakens relative to the level of the accumulated noise—the so-called signal-to-noise ratio. At some point along the facility, the signal weakens to the point that it must be amplified. The problem with amplification in analog systems is that the amplification process is nondiscriminatory—the amplifier amplifies whatever it is fed, which includes both the original signal and the noise it has accumulated along the way. The result of this is a louder, noisy signal, clearly not the most desirable outcome.

In digital systems, things happen a bit differently. Because digital transmission relies by its very nature on a limited set of discrete data points (zeroes and ones, or some combination thereof), it is easier for digital amplification equipment (usually called *repeaters* or *regenerators*) to identify the original signal and filter everything else out, resulting in a recreation of the original signal (hence, the term regenerator). We find, therefore, that digital networks are not cleaner than their analog counterparts; they're just better at dealing with the dirt.

Of course, there are myriad ways that analog systems can provide reliable transport services. Many coaxial cable systems are still analog today, but because the physical facility is shielded so well, errors rarely occur. By the same token, complex error correction and detection protocols have been developed that do a very good job of eliminating the problems of noise in analog transmission systems.

So now we understand the difference between signaling and content. Signaling describes the manner in which we represent the content that is being transported across the network, while content describes the information itself, which can be analog (such as voice) or digital (such as the stream of zeroes and ones that spew from the serial port of a personal computer).

TRADITIONAL ACCESS TECHNOLOGIES

Because the human voice only requires 4 KHz of bandwidth to achieve reasonable transmission quality, the analog telephone network, including the local loop, has been designed to allocate exactly that much bandwidth to each conversation. Since the late 1960s, the inner core of the network has been progressively digitized, starting with the introduction of T-Carrier in 1962 for central-office-to-central-office transport and the arrival of the first digital switches in the 1970s. The local loop, however, has largely remained analog, in spite of the efforts of such technologies as ISDN (discussed later) to make digital headway.

In 1981, IBM turned the world on its ear with the introduction of the PC, and in 1984, the Macintosh arrived, bringing computing power to the proverbial masses. Shortly thereafter, hobbyists and hackers[2] began to take advantage of emergent modem technology and created online databases—the first bulletin board systems that enabled people to send simple text messages to each other. This accelerated the modem market dramatically, and before long data became a common component of local loop traffic. Similarly, the business world found more and more applications for data, and the need to move that data from place to place also became a major contributor to the growth in data traffic on the world's telephone networks.

At this point, data did not represent a problem for the bandwidth-limited local loop. The digital information created by a computer and intended for transmission through the telephone network was received by a modem, converted into a modulated analog waveform that fell within the 4-KHz voice band, and fed to the network without incident. The modem's job was (and is) quite simple: When a computer is doing the talking, the modem must make the network think it is talking to a telephone. It did this by ensuring that the signals it transmitted emulated the signals a telephone might generate.

Over time, modem technology advanced, and local loop quality improved, enabling the local loop to provide higher and higher bandwidth. This increasing bandwidth was made possible through clever signaling schemes that enabled a single sig-

---

[2]It should be observed that the original hackers were not viewed as evil creatures, but rather as tinkerers who pushed the creativity envelope.

naling event to transport more than a single bit. Consider the following example. If we use amplitude modulation to represent digital data, we might have a system that uses a high-amplitude (loud) wave to represent a one and a low-amplitude wave (soft) to represent a zero. In this system, each signal represents a single bit, which means that if we are limited to the 4-KHz voice band, our bit rate is similarly limited. In fact, Harry Nyquist demonstrated that the fastest rate at which we can signal in a bandwidth-limited system is at a rate that is no faster than twice the highest frequency in the band within which we are operating. In the voice band, then, the fastest signaling rate we can ever achieve is 2×4 KHz, or 8,000 signals per second.

So why do we care about this esoterica? We care because the signaling rate has a direct impact on the achievable bit rate, and that is a central service component. If the width of the channel limits the signaling rate, then it obviously also limits the bits represented by those signals. So what are the available options? If the channel bandwidth limits the signaling rate, then obviously a bigger pipe means more signals. The problem is that bigger pipes mean more money because bandwidth is expensive. So although this may work, it is not necessarily the most desirable option. Besides, the voice band is limited to 4 KHz, and the switches are looking for information to come out of that limited channel. So a bigger voice band is not a viable option.

Alternatively, we might create a way to represent multiple bits on each transmitted signal. In fact, this is precisely what modern modems do. Consider our previous example in which we described a system where a single bit was represented by each discrete signal—high amplitude connoted a one, low amplitude a zero. What happens if we now add a little complexity? Instead of relying purely on amplitude modulation, we now add frequency modulation as well, as shown in Figure 1-4. As a consequence, we now have four definable states that we can use to represent information instead of two: high amplitude, high frequency; high amplitude, low frequency; low amplitude, high frequency; and low amplitude, low frequency. Instead of carrying a single bit, each of these can now carry two

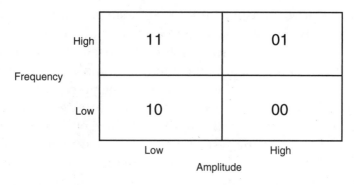

FIGURE 1-4   Multibit encoding.

bits as shown, which means that our bit rate is now twice our signaling rate, often referred to as *baud*.

Consider the consequences of this: If the signaling rate is limited by the bandwidth of the channel in which we are required to operate, we can overcome this limitation by convincing each signal to carry multiple bits, thus achieving higher bit rates in spite of a bandwidth-limited channel. There are many examples of this, ranging from the inner workings of 9600 bit-per-second modems that encode four bits per signal to cable systems capable of encoding as many as 10 bits per signal. Needless to say, at these encoding levels the degree to which noise is present can have a deleterious effect on the capability to cleanly transport data, so noise reduction techniques are of paramount importance.[3] Some modems employ highly sophisticated error detection and correction techniques such as trellis-coded modulation; other systems rely on the physical nature of the medium over which they transport their data cargo to reduce the error rate. Coaxial cable, with its inherent shielding, is one example of this; fiber, with its immunity to electrical interference, is another. Both media are still susceptible to errors, but far less so than traditional metallic conductors.

---

[3]Work performed by Claude Shannon, often referred to as Shannon's Law, concludes that the number of bits that can be encoded on a single signal is limited by the signal-to-noise ratio. In effect, the more noise, the less bits per signal.

The analog local loop is employed for a variety of voice and data applications today in both the business and residence marketplaces. The new lease on life it enjoys thanks to advanced modem technology as well as a focus by installation personnel on the need to build clean, reliable outside plant has resulted in the development of faster access technologies designed to operate across the analog local loop.

This does not mean to imply that traditional, preexisting technologies will disappear anytime soon. There are still hundreds of thousands, if not millions, of modems installed in computers today that operate at speeds in the range of 33.6 Kbps. New machines, however, are routinely equipped with V.90-based 56-Kbps modems and in some cases with ADSL and even cable modems. A discussion of these various access devices and the relative advantages and disadvantages they offer follows.

MARKETPLACE REALITIES

According to both private and government-conducted studies, there are approximately 60 million households today that host home office workers, including both telecommuters and those who are self-employed and work out of their homes. These workers require the ability to connect to remote LANs and corporate databases, retrieve e-mail, access the Web, and in some cases conduct videoconferences with colleagues and customers. The traditional local loop, with its bandwidth-limited capabilities, is not capable of easily satisfying these requirements with standard modem connectivity. Dedicated private line service, which would certainly solve the problem, is out of the financial reach of many of these businesses or is not enabled by parent companies for remote workers. Other solutions are required, and these have emerged in the form of a new flock of access technologies that take advantage of either a conversion to end-to-end digital connectivity (ISDN) or expanded capabilities of the traditional analog local loop [x-Type DSL (xDSL) and 56K modems]. In some cases, fixed wireless and satellite are causing excitement in the industry.

56-KBPS MODEMS

One of the most important words in telecommunications is virtual. It is used in a variety of ways, but in reality only has one meaning. If you see the word virtual associated with a technology or product, you should immediately say to yourself, it's a lie.

56-Kbps modems, governed by the *International Telecommunications Union* (ITU) standard called V.90, are good examples. They have garnered a significant amount of interest since they were introduced a few years ago. Under certain circumstances they do offer higher access speeds designed to satisfy the increasing demands of bandwidth-hungry applications and increasingly graphics-oriented Web pages. The problem with these so-called 56K modems is that they are virtual modems in that they do not really provide true 56K access, even under the best of circumstances.

These devices are designed to provide asymmetric bandwidth, with slightly less than 56 Kbps delivered downstream toward the customer, and even less bandwidth (33.6 Kbps) in the upstream direction. Although this may seem odd, it makes sense given the requirements of most applications today that require modem access. A Web session, for example, requires little bandwidth in the upstream direction to request that a page be downloaded; the page itself, however, may require significantly more because it may be replete with text, graphics, Java applets, and even small video clips. Because the majority of modem access today is for Internet surfing, this asymmetric technique is adequate for most users today. Tests show that users get approximately 20 percent greater throughput with a 56K modem than they get with 33.6 Kbps.

56K modem technology experienced something of a market rebellion when it was first introduced. Two industry players simultaneously introduced 56K standards that worked well but were incompatible with one another. Motorola, Rockwell, and Lucent collectively introduced their K56 Flex technology, while US Robotics (now 3Com) introduced a standard called X2. The problem with this simultaneous, incompatible product introduction was that it occurred at a time when online usage was at an all-time high and the marketplace—particularly the ISPs—were looking to buy enormous volumes of modems to augment their modem pools and slake their customers' hunger for greater bandwidth. Unfortunately, because there were two mutually incompatible standards, both introduced by reputable firms, the modem market came to a halt while would-be buyers waited for the standards battle to be won by one or the other. Ultimately, that did not happen; instead, the ITU in its wisdom issued an overlay standard called V.90 that hid the incompatibilities of each standard from the other, making the choice of one over the other a nonissue. V.90, mentioned earlier, is the dominant standard for 56-Kbps modems today.

The limitations of 56K modems stem from a number of factors. One of them is the fact that under current FCC regulations (Part 68), line voltage supplied to a communications facility is limited such that the maximum achievable bandwidth is 53 Kbps in the downstream direction. Another limitation is that these devices require that only a single analog-to-digital conversion occur between the two end points of the circuit. This typically occurs on the downstream side of the circuit and usually at the interface point where the local loop leaves the central office. Consequently, downstream traffic is less susceptible to the noise created during the analog-to-digital conversion process, while the upstream channel is affected by it and is therefore limited in terms of the maximum bandwidth it can provide. In effect, 56K modems, in order to achieve their maximum bandwidth, require that one end of the circuit, typically the central office end, be digital.

The good news with regard to 56K modems is that even in situations where the 56 Kbps speed is not achievable, the modem will fall back to whatever maximum speed it can fulfill. Furthermore, no premises wiring changes are required, and because this device is really nothing more than a faster modem, the average customer is comfortable with migration to the new technology. This is certainly demonstrated by sales volume; as was mentioned before, most PCs today are automatically shipped with a 56K modem.

JUST OUT: V.92

Even at 56 Kbps, V.90 modems had to put up with users grousing about the asymmetric nature of data transport supported by the modem (53 Kbps downstream, 33.6 Kbps upstream) and the fact that while connected, incoming calls were inaccessible. First announced in July 2000 and now becoming available, a new set of modems based on the V.92 standard is allaying those concerns. V.92 modems add three new capabilities that expand upon the capabilities offered by V.90 modems: Quick Connect, Modem-on-Hold, and PCM Upstream.

Many in the industry argued that the growing deployment of ISDN, DSL, and cable modems would obviate the need for another analog modem standard, even with its enhanced capabilities. As we will see a bit later in this chapter, neither of these relatively new technologies have enjoyed raging successes, the result of which is that most industry authorities believe that as many as 55 percent of Internet users will still be using analog modem access technologies in 2004. The need for V.92 is, therefore, justified.

The three added capabilities that V.92 brings to the modem table are significant. The first of these, Quick Connect, does precisely what the name implies: It reduces the typical connect time by 50 percent or more, shrinking the average 25-second connection to as little as 10 seconds. It does this in a rather clever fashion. Part of the modem handshake process involves an assessment of the characteristics of the transmission facility to determine the maximum speed at which the two modems can communicate. Most users make multiple calls to the same number over the same facility, even while traveling (from a hotel room, for example). The characteristics of the line rarely change between calls, so V.92 modems take advantage of this fact by memorizing the line characteristics to reduce the length of successive handshake procedures.

The second new service is called Modem-on-Hold. This feature enables a user to receive an incoming call while maintaining a previously established connection to the Internet. The service requires that the standard telco-provided call-waiting feature be available on the line, but this is a trivial issue. The service also works in reverse: A user can maintain a modem connection while initiating a voice call.

The third feature, PCM Upstream, addresses the complaint of asymmetric transport. PCM Upstream increases the upstream data rate from 33.6 Kbps to a maximum of 48 Kbps, a 30 percent increase. It has no effect on the download speed, but most user complaints have been with regard to upstream connectivity. Furthermore, a new compression standard, V.44, improves the service provided by traditional V.42bis compression by as much as 200 percent for compressible data streams. This results in an overall throughput of as high as 300 Kbps.

All of these features taken together result in a significant service improvement over V.90 modems, and given the deployment and service issues that cable, ISDN, and DSL have suffered through, analog modems may well be around longer than we currently believe. They do, after all, work and can be (within reason) universally deployed.

ISDN

ISDN has been the proverbial technological roller coaster since its arrival as a concept in the late 1960s. Internationally, it has enjoyed significant success as a true, digital local loop technology. In the United States, however, because of competing, incompatible hardware implementations, high cost, and marginal availability, its deployment has been spotty at best. In market areas where providers have made it available at reasonable prices, it has been quite successful. The good news is that since 1997, ISDN has experienced something of a phoenix-like resurgence that has occurred in lockstep with the growing demand for bandwidth. This growth, illustrated in Table 1-1,[4] will ultimately slow as other technologies, such as xDSL, begin

[4]*Telephony Magazine*, July 1998.

TABLE 1-1   ISDN Growth, 1997–2002

YEAR	BRI	PRI	TOTAL
1997	933K (+30%)	49K (+60%)	982K (+29%)
1998	1.25M (+35%)	79K (+60%)	1.229M (+36%)
1999	1.637M (+30%)	119K (+50%)	1.757M (+31%)
2000	2.047M (+25%)	167K(+40%)	2.214M (+26%)
2001	2.456M (+20%)	217K (+30%)	2.674M (+21%)
2002	2.824M (+15%)	261K (+15%)	3.086M (+15%)

to disrupt the curve. It is interesting to note, however, that this slowdown has been retarded by the failure of alternative broadband access technologies (DSL and cable modems) to succeed in the marketplace. ISDN has, therefore, enjoyed something of a usage renaissance in the last two years.

THE TECHNOLOGY

The typical non-ISDN local loop is analog. Voice traffic is carried from an analog telephone to the central office using a frequency-modulated carrier; once at the central office, the signal is usually digitized for transport within the digital network cloud. On the one hand, this is good because it means that there is a digital component to the overall transmission path. On the other hand, the loop is still analog, and as a result, the true promise of an end-to-end digital circuit cannot be realized.

In ISDN implementations, local switch interfaces must be modified to support a digital local loop. Instead of using analog frequency modulation to represent voice or data traffic carried over the local loop, ISDN digitizes the traffic at the origination point, either in the voice set itself or in an adjunct device known as a *terminal adapter* (TA). The digital local loop then uses *time division multiplexing* (TDM) to create multiple channels over which the digital information is transported and that provide for a wide variety of truly integrated services.

THE BASIC RATE INTERFACE (BRI)

There are two well-known implementations of ISDN. The most common (and the one intended primarily for residence and small business applications) is called the *Basic Rate Interface* (BRI). In BRI, the two-wire local loop supports a pair of 64-Kbps digital channels known as *B-Channels* as well as a 16-Kbps *D-Channel*, which is primarily used for signaling but can also be used by the customer for low-speed (up to 9.6 Kbps) packet data. The B-Channels can be used for voice and data, and in some implementations, can be bonded together to create a single 128-Kbps channel for videoconferencing or other higher bandwidth applications.

Figure 1-5 shows the layout of a typical ISDN BRI implementation, including the alphabetic reference points that identify the regions of the circuit and the generic devices that make up the BRI. In this diagram, the LE is the local exchange, or switch. The NT1 is the network termination device that serves as the demarcation point between the customer and the service provider; among other things, it converts the two-wire local loop to a four-wire interface on the customer's premises. The *terminal equipment, type 1* (TE1) is an ISDN-capable device, such as an ISDN telephone, while a *terminal equipment, type 2* (TE2) is a non-ISDN-capable device, such as a *plain old telephone service* (POTS) telephone. In the event that a TE2 is used, a TA must be inserted between the TE2 and the NT1 to perform analog-to-digital conversion and rate adaptation.

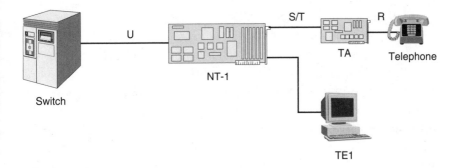

FIGURE 1-5   The ISDN Basic Rate Interface (BRI).

The reference points mentioned earlier identify circuit components between the functional devices just described. The U reference point is the local loop; the S/T reference point sits between the NT1 and the TEs; the R reference point is found between the TA and the TE2.

BRI APPLICATIONS

Although BRI does not offer the stunning bandwidth that other more recent technologies do, its bondable 64-Kbps channels provide reasonable capacity for many applications. The two most common today are the remote *local area network* (LAN) and Internet access. For the typical remote worker, the bandwidth available through BRI is more than adequate, and new video compression technology even puts reasonable quality videoconferencing within the grasp of the end user at an affordable price. 64 Kbps makes short shrift of LAN-based text file downloads and reduces the time required for graphics-intensive Web page downloads to reasonable levels—the World Wide Wait becomes a minor annoyance in the grand scheme of things.

Another application that has enjoyed the capabilities of ISDN is videoconferencing. Most of the major vendors of videoconferencing equipment have designed systems that have the capability to bind the two 64-Kbps B-Channels into a single 128-Kbps higher-speed channel, plenty of bandwidth for modern video *coder-decoders* (CODECs).

AO/DI

A relatively new arrival on the ISDN scene is an offering known as *Always On, Dynamic ISDN* (AO/DI). In AO/DI, the 16-Kbps D-Channel on the BRI, normally used for signaling, is recruited as a primary transport conduit for user data that normally employs a 64-Kbps B-Channel. If a user with AO/DI is connected to an e-mail service, for example, the packet-mode D-Channel can be used to deliver messages from the e-mail server to the user without requiring the services of a B-Channel.

Should higher bandwidth be required to download a large attachment, view a Web page, or engage a more bandwidth-hungry application, a B-Channel can be activated, used, and torn down when it is no longer needed (see Figure 1-6). This technique is easier on the network because it makes more intelligent use of available bandwidth and therefore places less of a burden on the switch. It does, of course, require the customer to have an account with a packet services provider to handle the D-Channel packet traffic, but because these services are generally billed on a volume-of-packets-transmitted basis, the cost is relatively low. Most users who employ AO/DI recognize that the cost to have the packet service is far outweighed by the instantaneous access to online services that the technology provides.

THE PRIMARY RATE INTERFACE (PRI)

The other major implementation of ISDN is called the *Primary Rate Interface* (PRI). The PRI is really nothing more than a standard T- or E-Carrier in that it is a four-wire circuit (local loop), uses AMI and B8ZS for ones-density control and signaling, and sports a complement of 24 or 30 64-Kbps channels that can be distributed among a collection of users as the

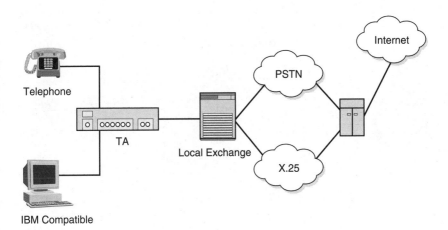

FIGURE 1-6 Always-On/Dynamic ISDN.

customer sees fit (see Figure 1-7). A T-Carrier, often called a T-1, is a 1.544-Mbps circuit that provides a total of 24 channels to the user, each of them 64 Kbps and known as a *DS-0*. Originally designed as an intra-office trunking scheme, it was believed that there would never be a reason why a customer would ever have to know of the existence of the T-Carrier, given the vast amounts of bandwidth it proffered. Today, the T-Carrier is routinely deployed as a medium-bandwidth loop solution to the customer business premise for both voice and data applications. Outside the United States and a handful of other countries, the T-Carrier is not used; most countries use the E-Carrier, a 32-channel system that offers 30 64-Kbps channels to the customer while reserving one channel each for signaling and synchronization. Functionally, the T- and E-Carriers are largely the same from a service delivery perspective.

In PRI, the signaling channel operates at 64 Kbps (unlike the 16-Kbps D-Channel in the BRI) and is not accessible by the user. It is used solely for signaling purposes—that is, it cannot carry user data. The primary reason for this is service protection. In the PRI, the D-Channel is used to control the goings-on of 23 B-Channels and therefore requires significantly more bandwidth than the BRI D-Channel. Furthermore, the PRI standards enable multiple PRIs to share a single D-Channel, which makes the D-Channel's operational consistency all the more critical.

The functional devices and reference points are not appreciably different from those of the BRI. Here, the local loop is a

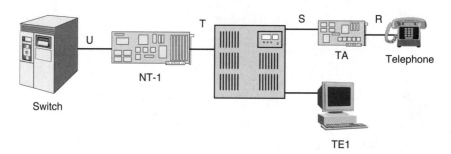

FIGURE 1-7   The ISDN Primary Rate Interface (PRI).

four-wire T-carrier rather than a traditional two-wire local loop, but it is still identified as the U reference point. In addition to an NT1, we now add an NT2, which is a service distribution device, usually a *private branch exchange* (PBX), which allocates the PRI's 24 channels to customers. The S/T reference point is now divided; the S reference point sits between the NT2 and TEs, while the T reference point is found between the NT1 and the NT2.

PRI service also has the capability to cluster B-Channels into super-rate channels to satisfy the bandwidth requirements of higher bit rate services such as medical imaging and video-conferencing. These clusters of B-Channels are called *H-Channels* and are provisioned as shown in Table 1-2.

PBX APPLICATIONS

The PRI's marketplace is the business community, and its primary advantage is pair gain—that is, it conserves copper pairs by multiplexing the traffic from multiple user channels onto a shared, four-wire circuit. Inasmuch as a PRI can deliver the equivalent of a minimum of 23 voice channels to a location over a single circuit, it is an ideal technology for a number of applications, including the interconnection of a PBX to a local switch, dynamic bandwidth allocation for higher-end videoconferencing applications, and the interconnection between an ISP's network and that of the local telephone company.

A PBX is really nothing more than a remote switch that resides on the customer's premise and is connected to the

TABLE 1-2   Primary Rate Channel Bandwidth

CHANNEL	BANDWIDTH
H0	384 Kbps (6B)
H10	1.472 Mbps (23B)
H11	1.536 Mbps (24B)
H12	1.920 Mbps (30B)

service provider's switch in the central office by one or more carrier facilities, typically T-Carriers, as shown in Figure 1-8. The PBX enables the IT and voice service managers of a large corporate location to self-manage their telecomm services and offered features. For example, the PBX enables a corporation to implement four- or five-digit dialing between users that are behind the PBX. The dial tone they hear is actually provided by the PBX, not the PSTN, unless they want to make a call over the public network. In that case, they must first dial 9, which instructs the PBX to pass the call on to the PSTN by bringing up a foreign dial tone (from the service provider) and handing off the call.

Some PBXs are ISDN-capable on the line (customer) side, meaning that they have the capability to deliver ISDN services to users that emulate the services that would be provided over a direct connection to an ISDN-provisioned local loop. On the trunk (switch) side, the PBX is connected to the local switch via one or more T-Carriers (T1s), which in turn provide access to the telephone network. This arrangement results in significant savings, faster call setup, more flexible administration of trunk resources, and the capability to offer a diversity of services through the granular allocation of bandwidth as required.

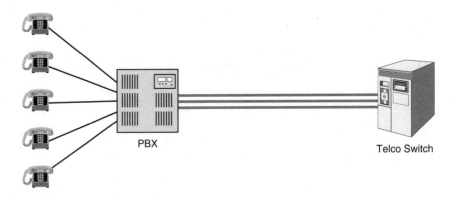

FIGURE 1-8   A typical PBX installation.

VIDEOCONFERENCING

As we noted earlier, videoconferencing, an application that now enjoys widespread acceptance in the market thanks to affordable, effective video CODECs, has moved from the boardroom to the home office. While BRI provides adequate bandwidth for casual conferencing on a reduced-size PC screen, PRI is required for high-quality, television-scale sessions. Its capability to provision bandwidth on demand for such applications makes it an ideal solution. All major videoconferencing equipment manufacturers have embraced PRI as an effective connectivity solution for their products and have consequently designed their products in accordance with accepted international standards—specifically, H.261 and H.320.

AUTOMATIC CALL DISTRIBUTION (ACD)

Another significant application for PRI is call routing in the call center environment. Ultimately, call centers represent critical decision points within a corporate environment, and the degree to which they are successful at what they do translates directly into visible manifestations of customer service. If calls are routed quickly and accurately based on some well-designed internal decision-making process, customers are happy, and the call center provides an effective image of the corporation and its services before the customer.

Automatic Call Distribution (ACD) is a popular switch feature that enables a customer to create custom routing tables in the switch so that incoming calls can be handled most effectively on a call-by-call basis. The ACD feature often relies on information culled from both corporate sources and the SS7 network's SCP databases, but the ultimate goal is to provide what appears to each caller to be customized answering. For example, some large call centers handle incoming calls from a variety of countries and therefore potentially different language groups. Using caller ID information, an ACD can filter calls as they arrive and route them to language-appropriate operators. Similarly, the ACD can change the assignment of voice channels on a demand basis. For example, if the call center is large and receives calls from multiple time zones simultaneously, there may be a need to add incoming trunks to one region of the call center based on time-of-day call volumes, while reducing the total coverage in another. By reallocating the 64K channels of the PRI(s), bandwidth can be made available as required. Furthermore, it can be done automatically, triggered by time of day or some other pre-identified event.

IN SUPPORT OF THE REMOTE AGENT

Many companies now support distributed call centers. In this model, agents are not necessarily physically collocated but may in fact be working in their homes. Each remote agent has a BRI line to their location that provides them with two channels— one for voice, the other for access to database information, and whatever applications they must use to perform their jobs. The ACD feature automatically routes calls to each agent according to some predefined set of functions. The incoming calls appear on one B-channel; the other is reserved for database access.

Both corporations and government agencies alike have applauded this model because it supports the basic tenets of telecommuting. Employees who work this way are happy because they can stay home to care for children or elderly parents and are generally (according to most studies that have been conducted) much more productive than those who work in a traditional office setting. The government is happy as well, because telecommuting reduces the total number of cars on the road, thus supporting federal clean air mandates. Corporations benefit because real estate costs can be reduced; if they legitimately support the program, they can provision statistically less space for employees and implement hoteling philosophies. Hoteling is a relatively new technique for office space management in which employees, particularly those who travel for a significant proportion of their professional time, do not have dedicated cubicle or office space. Instead, their personal effects are stored in rollaround cabinets that can easily be moved from one place to another. On those occasions when a road warrior employee must work in the office, they are dynamically assigned a cubicle, to which they trundle their personal cabinet. Once there, they call a number that causes the switch to auto-matically assign their office phone number to the phone line in that cubicle. As one employee told me who works in a facility that uses hoteling, "It took me a while to get used to the fact that I no longer had a dedicated office, but once I got accus-tomed to the idea, it was actually a good thing. It made me much more flexible, and made me much less dependent on 'space' as a part of my success in the workplace."

Ironically, ISDN is the earliest example of true technological convergence. The value of ISDN does not lie in the fact that it enables multiple information streams to share a common phys-ical transmission channel—that capability has been available since the turn of the century when scientists first invented har-monic telegraphs. Its value lies in the fact that it enables the *logical integration of services* over a shared channel, the essence of the convergence phenomenon. Although it does not offer the high bandwidth provided by competitive wireline technologies, such as cable modems and xDSL, it has the advantage of being able to offer a full suite of capabilities, including reasonable bandwidth for quite a few applications, *right now*—a claim that alternative technologies can't necessarily make.

T-CARRIER AND VOICE DIGITIZATION

The voice network, including both transmission facilities and switching components, was exclusively analog in nature until 1962, when T-Carrier emerged as an intra-office trunking scheme. The technology was originally introduced as a short-haul, four-wire facility to serve metropolitan areas. Over the years, it evolved to include coaxial cable facilities, digital microwave systems, fiber, and satellite, and today is a com-monly deployed access technology for medium-speed applications. When it was first introduced, there was no concept that the outside world (that is, outside the walls of the central office) would ever have to be aware of its existence. After all, a customer would never require the almost unimaginable bandwidth that T-Carrier provided!

As the network topology improved, so, too, did the switching infrastructure. In 1976, AT&T introduced the 4ESS switch primarily for toll applications and followed it up with the 5ESS in 1981 for local switching access as well as a variety of remote switching capabilities. Nortel, Siemens, and Ericsson all followed suit with equally capable hardware. In 1983, the first tariff for T-1 was released, and the service was on its way to becoming mainstream.[5]

The goal of digitizing the human voice for transport across an all-digital network grew out of work performed at Bell Laboratories shortly after the turn of the century. That work led to a discrete understanding of not only the biological nature and spectral makeup of the human voice, but also to a better understanding of language, sound patterns, and the sounded emphases that comprise spoken language.

THE NATURE OF VOICE

A typical voice signal comprises frequencies that range from approximately 30 Hz to 20 KHz. Most of the speech energy, however, lies between 300 Hz and 3,300 Hz, the so-called voice band. Early experiments showed that the frequencies below 1 KHz provide the bulk of recognizability and intelligibility, while the higher frequencies provide richness, articulation, and natural sound to the transmitted signal.

As we noted earlier, the human voice comprises a remarkably rich mix of frequencies, and this richness comes at a considerable price. In order for telephone networks to transmit

---

[5]It seems that there are as many explanations for the "T" in T-Carrier as there are circuit miles of the service deployed. Let us dispense with the rumors: The T stands for "terrestrial."

voice's entire spectrum of frequencies, significant network bandwidth must be made available to every ongoing conversation. There is a substantial price tag attached to bandwidth; it is a finite commodity within the network, and the more of it that is consumed, the more it costs.

Thankfully, work performed at Bell Laboratories at the beginning of the twentieth century helped network designers confront this challenge head-on. When the telephone network first began to spread its tentacles across the continent, there were no switches. Initially, subscribers in towns bought individual phones and phone lines to each person they had a need to speak with, resulting in the famous photographs (see Figure 1-9) of metropolitan telephone poles with tier upon tier of cross-pieces, festooned with aerial wire, and its mathematical representation: $n(n - 1)/2$, the total number of circuits that would be required for n people to be able to speak with everyone else in the community, as shown in Figure 1-10. For example, if 100 people in a neighborhood wanted to be able to speak with each other, they would require the phone company to install a total of $(100)*(99)/2$, or 4,950 circuits—for 100

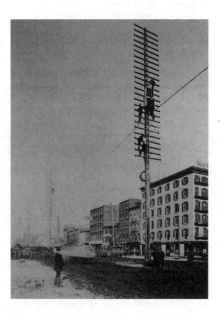

FIGURE 1-9   Crossbars on early outside plant installation.

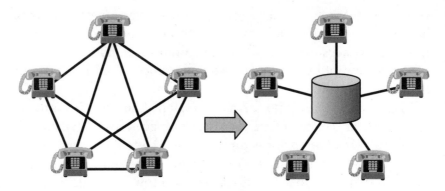

FIGURE 1-10 $n(n-1)/2$ in action: From 10 circuits to 5.

people. Clearly, this was economically out of the question, not to mention the fact that the quantity of aerial cable threatened to block out the sun, potentially precipitating the next ice age and the end of civilization as we know it.

The solution to this quandary came in two forms: central office switches and multiplexing. Central office switches enabled the phone company to provide the same level of connectivity, but required each customer to have only *one* circuit. The responsibility for setting up the connection to the called party resided with the switch in the central office, rather than with the customer. The first switches, of course, were operators ("Sarah, get me Andy over at the courthouse!"); true mechanical switches didn't arrive until 1982, when Almon Strowger's step-by-step switch was first installed by Automatic Electric.

Equally important was the concept of multiplexing, which enabled multiple conversations to be carried simultaneously across a single shared physical circuit. The first systems used *frequency-division multiplexing* (FDM), a technique made possible by the development of the vacuum tube, in which the range of available frequencies is divided into chunks that are then parceled out to subscribers. For example (and this is *only* an example), subscriber 1 might be assigned the range of frequencies between 0 and 4,000 Hz, while subscriber 2 is assigned 4,000 to 8,000 Hz, 3 has 8,000 to 12,000 Hz, and so on, up to the maximum range of frequencies available in the system, as illustrated in Figure 1-11. In FDM, we often observe that users are given some of the frequency all of the time, meaning that they are free to use their assigned frequency allocation at any time, but may *not* step outside the bounds. Early FDM systems were capable of transporting 24–4 KHz channels for an overall system bandwidth of 96 KHz. FDM, although largely replaced today by more efficient systems that will be discussed later, is still used in cellular telephone and microwave systems, among others.

This model worked well in the early telephone systems. Because the lower regions of the 300- to 3,300-Hz voice band carry the frequency components that enable recognizability and intelligibility, telephone engineers concluded that although the higher frequencies enrich the transmitted voice, they are not necessary for both parties to recognize and understand each other. This understanding of the makeup of the human voice

FIGURE 1-11 Analog frequency assignments in the traditional telephone network.

helped them create a network that is capable of faithfully repro-
ducing the sounds of a conversation while keeping the cost of
consumed bandwidth to a minimum. Instead of assigning the
full complement of 20 KHz to each end of a conversation, they
employed filters to bandwidth-limit each user to approximately
4,000 Hz, a resource savings of some 60 percent. Within the
network, subscribers were FDMed across shared physical facilities, thus enabling the telephone company to efficiently con-
serve network bandwidth.

As with any technology, there were downsides to FDM. It
is an analog technology and therefore suffers from the shortcomings that have historically plagued all transmission systems. The wire over which information is transmitted behaves
like a longwire antenna, picking up noise along the length of
the transmission path and very effectively homogenizing it
with the voice signal. Additionally, the power of the transmitted signal diminishes over distance, and if the distance is great
enough, the signal will have to be amplified to make it intelligible at the receiving end. Unfortunately, the amplifiers used
in the network are not particularly discriminating. They have
no way of separating the voice wheat from the noise chaff, as
it were; the result is that they convert a weak, noisy signal into
a loud noisy signal—better, but far from ideal. A better solution was needed.

The better solution came about with the development of
TDM, which became possible with the development of the tran-
sistor and integrated circuit electronics. TDM is a digital transmission scheme, which implies a small number of discrete
signal states rather than the essentially infinite range of values
employed in analog systems. Although digital systems are as
susceptible to noise impairment as their analog counterparts,
the discrete nature of their binary signaling makes it relatively
easy to separate the noise from the transmitted signal. In T-
Carrier systems, there are only three valid signal values: one
positive, one negative, and zero; anything else is construed to be
noise. It is, therefore, a trivial exercise for digital repeaters (the
digital equivalent of an analog amplifier) to discern what is

desirable and what is not, thus eliminating the problem of cumulative noise.

Because of their rich frequency content, digital signals require a broad transmission channel—they cannot be transmitted across the bandwidth-limited channels of the traditional telephone network. In digital T-Carrier facilities, the equipment that restricts the individual transmission channels to 4 KHz chunks is removed, thus giving each user access to the full breadth of available spectrum across the shared physical medium. In FDM systems, we observed that we give users some of the frequency all of the time; in TDM systems, we turn that around and give users *all* of the frequency *some* of the time. See Figure 1-12.

Digitization brings with it a cadre of advantages, including improved voice and data transmission quality, better maintenance and troubleshooting capabilities, and therefore reliability, and dramatic improvements in configuration flexibility. In T-Carrier systems, the TDM is known as a *channel bank*; under normal circumstances, it enables 24 circuits to share a single, 4-wire facility.

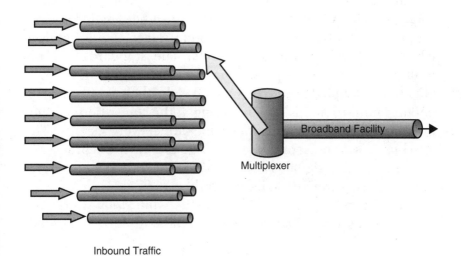

Inbound Traffic

FIGURE 1-12   Time division multiplexing.

VOICE DIGITIZATION

The process of converting analog voice to a digital representation in the modern network is a logical and straightforward process. It comprises four distinct steps: *pulse amplitude modulation* (PAM) sampling, in which the amplitude of the incoming analog waveform is sampled every 125 microseconds; companding, during which the values are weighted toward those most receptive to the human ear; quantization, in which the weighted samples are given values on a nonlinear scale; and, finally, encoding, in which each value is assigned a distinct binary value. Each of these stages of *pulse code modulation* (PCM) will now be discussed in detail.

PCM

Thanks to work performed by Harry Nyquist at Bell Laboratories in the 1920s, we know that to optimally represent an analog signal as a digitally encoded bitstream, the analog signal must be sampled at a rate that is equal to twice the bandwidth of the channel over which the signal is to be transmitted. Because each analog voice channel is allocated 4 KHz of bandwidth, it follows that each voice signal must be sampled at twice that rate, or 8,000 samples per second. In fact, that is precisely what happens in T-Carrier systems. The standard T-Carrier multiplexer accepts inputs from 24 analog channels. Each channel is sampled in turn, every one eight-thousandth of a second in round-robin fashion, resulting in the generation of 8,000 pulse amplitude samples from each channel every second.[6] This PAM process represents the first stage of PCM, the process by which an analog signal is converted to a digital signal for transmission across the T-Carrier network.

The second stage of PCM is called *quantization*. In quantization, we assign values to each sample within a constrained

---

[6]The sampling rate is important for several reasons. If the sampling rate is too high, too much information is transmitted, and bandwidth is wasted; if the sampling rate is too low, then we run the risk of *aliasing*. Aliasing is the interpretation of the sample points as a false waveform, due to the paucity of samples.

range. For illustration purposes, imagine what we now have before us. We have replaced the continuous analog waveform of the signal with a series of amplitude samples that are close enough together that we can discern the shape of the original wave from their collective amplitudes. Imagine also that we have graphed these samples in such a way that the wave of sample points meanders above and below an established zero point on the x-axis, so that some of the samples have positive values and others are negative, as shown in Figure 1-13.

The amplitude levels enable us to assign values to each of the PAM samples, although a glaring problem with this technique should be obvious to the careful reader. Very few of the samples actually line up exactly with the amplitudes delineated by the graphing process. In fact, most of them fall *between* the values, as shown in Figure 1-14. It doesn't take much of an intuitive leap to see that several of the different samples will be

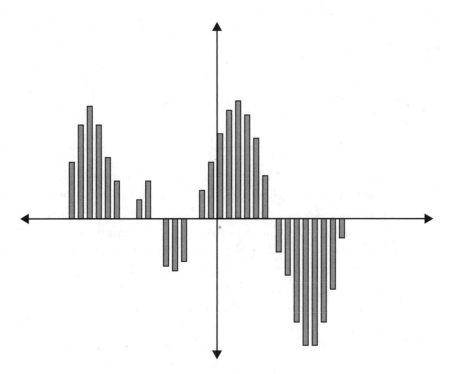

FIGURE 1-13   PAM samples.

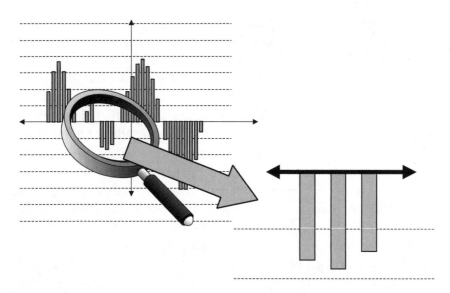

FIGURE 1-14   Quantizing the samples.

assigned the same digital value by the CODEC that performs this function, yet they are clearly *not* the same amplitude. This inaccuracy in the measurement method results in a problem known as *quantizing noise* and is inevitable when linear measurement systems, such as the one suggested by the drawing, are employed in CODECs.

Needless to say, design engineers recognized this problem rather quickly and equally quickly came up with an adequate solution. It is a fairly well-known fact among psycholinguists and speech therapists that the human ear is far more sensitive to discrete changes in amplitude at low-volume levels than it is at high-volume levels, a fact not missed by the network designers tasked with optimizing the performance of digital carrier systems intended for voice transport. Instead of using a linear scale for digitally encoding the PAM samples, they designed and employed a nonlinear scale that is weighted with much more granularity at low-volume levels—that is, close to the zero line —than at the higher-amplitude levels, as shown in Figure 1-15. In other words, the values are extremely close together near the x-axis and get farther and farther apart as they travel up and

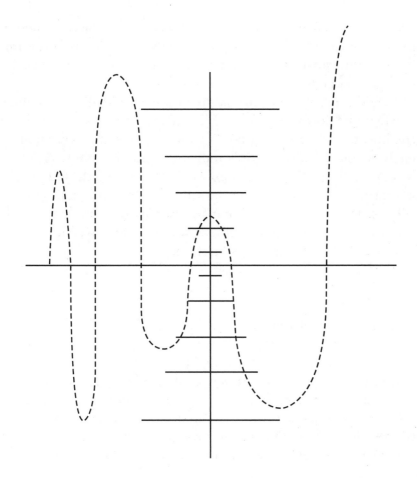

FIGURE 1-15 Companding the voice signal.

down the y-axis. This nonlinear approach keeps the quantizing noise to a minimum at the low-amplitude levels where hearing sensitivity is the highest and enables the noise to creep up at the higher amplitudes, where the human ear is less sensitive to its presence. It turns out that this is not a problem because the inherent shortcomings of the mechanical equipment (microphones, speakers, and the circuit itself) introduce slight distortions at high-amplitude levels that hide the effect of the nonlinear quantizing scale.

This technique of compressing the values of the PAM samples to make them fit the nonlinear quantizing scale results in a bandwidth savings of more than 30 percent. In fact, the actual process is called companding because the sample is first compressed for transmission and then expanded for reception at the far end, hence the term.

The actual graph scale is divided into 255 distinct values above and below the zero line.[7] There are eight segments above the line and eight below (one of which is the shared zero point); each segment, in turn, is subdivided into 16 steps. A bit of binary mathematics now enables us to convert the quantized amplitude samples into an 8-bit value for transmission. For the sake of demonstration, let's consider a positive sample that falls into the eleventh step in segment seven. The conversion would take on the following representation:

0 111 1011

The initial 0 indicates a positive sample, 111 indicates the seventh segment, and 1011 indicates the eleventh step in the segment. We now have an 8-bit representation of an analog amplitude sample that can be transmitted across a digital network, and then be reconstructed with its many counterparts as an accurate representation of the original analog waveform at the receiving end. This entire process is known as PCM, and the result of its efforts is often referred to as *toll-quality voice*.

ALTERNATIVE DIGITIZATION TECHNIQUES

Although PCM is perhaps the best-known, high-quality voice digitization process, it is by no means the only one. Advances in *Complementary Metal-Oxide Semiconductor* (CMOS) architecture and improvements in the overall quality of the telephone network have made it possible for encoding schemes to be

---

[7]In the United States and a handful of other countries, including Canada and Japan, the encoding scheme is known as μ-Law (Mu-Law); the rest of the world relies on a slightly different standard known as A-Law.

developed that use far less bandwidth than traditional PCM. For the corporation, this means a significant savings in bandwidth and therefore cost. In this next section, we will consider some of these techniques.

ADAPTIVE DIFFERENTIAL PULSE CODE MODULATION (ADPCM)

ADPCM is a technique that enables toll-quality voice signals to be encoded at a half-rate (32 Kbps) for transmission. ADPCM relies on the predictability that is inherent in human speech to reduce the amount of information required. The technique still relies on PCM encoding, but adds an additional step to carry out its task. The 64-Kbps PCM-encoded signal is fed into an ADPCM transcoder, which considers the *prior* behavior of the incoming stream to create a prediction of the behavior of the *next* sample. Here's where the magic happens: Instead of transmitting the actual value of the predicted sample, it encodes in four bits and transmits the *difference* between the actual and predicted samples. Because the difference from sample to sample is typically quite small, the results are generally considered to be very close to toll-quality. This 4-bit transcoding process, which is based on the known behavior characteristics of human voice, enables the system to transmit 8,000 4-bit samples per second, thus reducing the overall bandwidth requirement from 64 Kbps to 32 Kbps.[8]

CONTINUOUSLY VARIABLE SLOPE DELTA (CVSD)

CVSD is a unique form of voice encoding that relies on the individual values of individual bits to predict the behavior of the

---

[8]It should be noted that ADPCM works well for voice because the encoding and predictive algorithms are based upon its behavior characteristics. It does not, however, work as well for higher bit rate data (above 4800 bps), which has an entirely different set of behavior characteristics.

incoming signal. Instead of transmitting the volume (height or y-value) of PAM samples, CVSD transmits information that measures the changing slope of the waveform. So instead of transmitting the actual change itself, it transmits the *rate* of change.

To perform its task, CVSD uses a reference voltage to which it compares all incoming values. If the incoming signal value is less than the reference voltage, then the CVSD encoder reduces the slope of the curve to make its approximation better mirror the slope of the actual signal. If the incoming value is *more* than the reference value, then the encoder will increase the slope of the output signal, again causing it to approach and therefore mirror the slope of the actual signal. With each recurring sample and comparison, the step function can be increased or decreased as required. For example, if the signal is increasing rapidly, then the steps are increased one after the other in a form of step function by the encoding algorithm. Obviously, the reproduced signal is not a particularly exact representation of the input signal. In practice, it is pretty jagged. Filters, therefore, are used to smooth the transitions.

CVSD is typically implemented at 32 Kbps, although it can be implemented at rates as low as 9600 bps. At 16 to 24 Kbps, recognizability is still possible; below 9600, recognizability is seriously affected, although intelligibility is not.

LINEAR PREDICTIVE CODING (LPC)

We mention LPC here only because it has carved out a niche for itself in certain voice-related applications, such as voice mail systems, automobiles, aviation, and electronic games that speak to children. LPC is a complex process, implemented completely in silicon, which enables voice to be encoded at rates as low as 2400 bps. The result is far from toll-quality, but is certainly intelligible, and its low bit rate capability gives it a distinct advantage over other systems.

LPC relies on the fact that each sound created by the human voice has unique attributes, such as frequency range, resonance, and loudness, among others. When voice samples are created in LPC, these attributes are used to generate prediction coefficients. These predictive coefficients represent linear combinations of previous samples, hence the name, *linear predictive coding*.

Prediction coefficients are created by taking advantage of the known *formants* of speech, which are the resonant characteristics of the mouth and throat, which give speech its characteristic timbre and sound. This sound, referred to by speech pathologists as the buzz, can be described by both its pitch and its intensity. LPC, therefore, models the behavior of the vocal cords and the vocal tract itself.

To create the digitized voice samples, the buzz is passed through an inverse filter that is selected based upon the value of the coefficients. The remaining signal, after the buzz has been removed, is called the *residue*.

In the most common form of LPC, the residue is encoded as either a *voiced* or *unvoiced* sound. Voiced sounds are those that require vocal cord vibration, such as the *g* in *glare*, the *b* in *boy*, and the *d* and *g* in *dog*. Unvoiced sounds require no vocal cord vibration, such as the *h* in *how*, the *sh* in *shoe*, and the *f* in *frog*. The transmitter creates and sends the prediction coefficients, which include measures of pitch, intensity, and whatever voiced and unvoiced coefficients that are required. The receiver undoes the process: it converts the voice residue, pitch, and intensity coefficients into a representation of the source signal, using a filter similar to the one used by the transmitter to synthesize the original signal.

DIGITAL SPEECH INTERPOLATION (DSI)

Human speech has many measurable (and therefore predictable) characteristics, one of which is a tendency to have embedded pauses. As a rule, people do not spew out a series of uninterrupted sounds; they tend to pause for emphasis, to collect their thoughts, to reword a phrase while the other person listens quietly on the end of the line. When speech technicians monitor these pauses, they discover that during considerably more than half of the total connect time, the line is silent.

DSI takes advantage of this characteristic silence to drastically reduce the bandwidth required for a single channel. Whereas 24 channels can be transported over a typical T-1 facility, DSI enables as many as 120 conversations to be carried over the same circuit. The format is proprietary and requires the setting aside of a certain amount of bandwidth for overhead.

A form of statistical multiplexing lies at the heart of DSI's functionality. Standard T-Carrier is a TDM scheme, in which channel ownership is assured: A user assigned to channel three will *always* own channel three, regardless of whether they are actually using the line. In DSI, channels are not owned. Instead, large numbers of users share a pool of available channels. When a user starts to talk, the DSI system assigns an available timeslot to that user and notifies the receiving end of the assignment. This system works well when the number of users is large because statistical probabilities are more accurate and indicative of behavior in larger populations than in smaller ones.

There is a downside to DSI, of course, and it comes in several forms. *Competitive clipping* occurs when more people start to talk than there are available channels, resulting in someone being unable to talk. *Connection clipping* occurs when the receiving end fails to learn what channel a conversation has been assigned within a reasonable amount of time, resulting in signal loss. Two approaches have been created to address these problems. In the case of competitive clipping, the system intentionally clips off the front end of the initial word of the second person who speaks. This technique is not optimal, but does prevent loss of the conversation and also obviates the problem of clipping out the middle of a conversation, which would be more difficult for the speakers to recover from. The loss of an initial syllable or two can be mentally reconstructed far more easily than sounds in the middle of a sentence.

A second technique used to recover from clipping problems is to temporarily reduce the encoding rate. The typical encoding rate for DSI is 32 Kbps; in certain situations, the encoding rate may be reduced to 24 Kbps, thus freeing up bandwidth for additional channels.

Both techniques are widely utilized in DSI systems.

FRAMING AND FORMATTING IN T-1

The standard T-Carrier multiplexer accepts inputs from 24 sources, converts the inputs to PCM bytes, and then TDMs the samples over a shared 4-wire facility. Each of the 24 input channels yields an 8-bit sample, in round-robin fashion, once every 125 microseconds (8,000 times per second). This yields an overall bit rate of 64 Kbps for each channel (8 bits per sample $\times$ 8,000 samples per second). The multiplexer gathers one 8-bit sample from each of the 24 channels and aggregates them into a 192-bit frame. To the frame it adds a frame bit, which expands the frame to a 193-bit entity. The frame bit is used for a variety of purposes that will be discussed in a moment.

The 193-bit frames of data are transmitted across the 4-wire facility at the standard rate of 8,000 frames per second for an overall T-1 bit rate of 1.544 Mbps. Keep in mind that 8 Kbps of the bandwidth consist of frame bits; only 1.536 Mbps belong to the user.

The earliest T-Carrier equipment was referred to as a D1 channel bank and was considerably more rudimentary in function than modern systems. In D1, every 8-bit sample carried seven bits of user information (bits one through seven) and one bit for signaling (bit eight), as shown in Figure 1-16. The signaling bits were used for exactly that: indications of the status of the line (on-hook, off-hook, busy, high and dry, and so on), while the seven user bits carried encoded voice information. Because only seven of the eight bits were available to the user, the result was considered to be less than toll quality (128 possible values, rather than 256). The frame bits, which in modern systems indicate the beginning of the next 192-bit frame of data, toggled back and forth between zero and one.

As time went on and the stability of network components improved, an improvement on D1 was sought after and found.

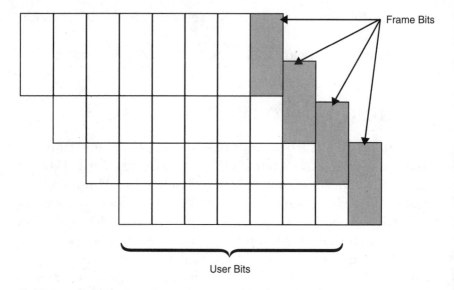

FIGURE 1-16   D1 framing.

Several options were developed, but the winner emerged in the form of the D4 or superframe format, shown in Figure 1-17. Rather than treat a single 193-bit frame as the transmission entity, superframe gangs together 12 193-bit frames into a 2,316-bit entity that obviously includes 12 frame bits. Please note that the bit rate has not changed; we have simply changed our view of what constitutes a frame.

Because we now have a single (albeit large) frame, we clearly don't need 12 frame bits to frame it; consequently, some of them can be redeployed for other functions. In superframe, six of the odd-numbered frame bits are referred to as *terminal framing bits* and are used to synchronize the channel bank equipment. The other six framing bits are called *signal framing bits* and are used indicate to the receiving device *where* robbed-bit signaling occurs.

In D1, the system reserved one bit from every sample for its own signaling purposes, which succeeded in reducing the user's overall throughput. In D4, that is no longer necessary. Instead, we signal less frequently and only occasionally rob a bit from the user. In fact, because the system operates at a high trans-

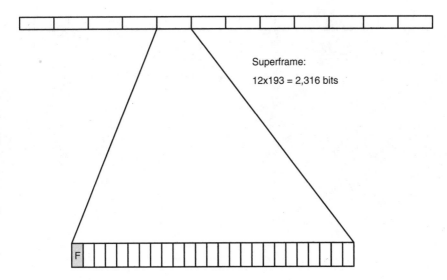

FIGURE 1-17   Superframe (D4) framing.

mission speed, network designers determined that signaling can occur relatively infrequently and still convey adequate information to the network. Consequently, bits are robbed from the sixth and eighth iteration of each channel's samples and then only the least significant bit from each sample. The resulting change in voice quality is negligible.

Back to the signal framing bits: Within a transmitted superframe, the second and fourth signal framing bits would be the same, but the sixth would toggle to the opposite value, indicating to the receiving equipment that the samples in that subframe of the superframe should be checked for signaling state changes. The eighth and tenth signal framing bits would stay the same as the sixth, but would toggle back to the opposite value once again in the twelfth, indicating once again that the samples in that subframe should be checked for signaling state changes.

Although superframe continues to be widely utilized, an improvement came about in the 1980s in the form of *extended superframe,* shown in Figure 1-18. ESF groups 24 frames into an entity instead of 12 and, like superframe, reuses some of the

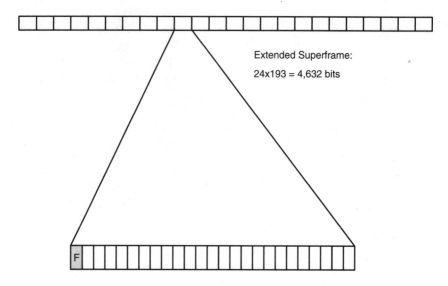

FIGURE 1-18   Extended Superframe (ESF) framing.

frame bits for other purposes. Bits 4, 8, 12, 16, 20, and 24 are used for framing and form a constantly repeating pattern (001011 . . .). Bits 2, 6, 10, 14, 18, and 22 are used as a 6-bit *cyclic redundancy check* (CRC) to check for bit errors on the facility. Finally, the remaining bits—all of the odd frame bits in the frame—are used as a 4-Kbps facility data link for end-to-end diagnostics and NM tasks. In this regard, ESF provides a major benefit over its predecessors. In earlier systems, if a customer reported trouble on the span, the span would have to be taken out of service for testing. With ESF, that is no longer necessary because of the added functionality provided by the CRC and the facility data link. The 4-Kbps *Embedded Operations Channel* (EOC), which results from the efficient reuse of framing bits in ESF, provides a dedicated facility between the *channel service units* (CSUs) on a T-carrier circuit over which the service provider can transmit diagnostic signals without affecting the customer.

The T-1 framing and transmission standard is used in North America and Japan, while the remainder of the world uses what is known as the CEPT E-1 standard. The *European Council on Post and Telecommunications Administrations* (CEPT) has largely been replaced today by the *European Telecommunica-*

tions Standards Institute (ETSI), but the name is still occasionally used.

E-1 differs from T-1 on several key points. First, it boasts a 2.048-Mbps facility rather than the 1.544-Mbps facility found in T-1. Second, it utilizes a 32-channel frame rather than 24. Channel one contains framing information and a *four-bit cyclic redundancy check* (CRC-4); Channel 16 contains all signaling information for the frame, and channels 1 through 15 and 17 through 31 transport user traffic. E-1 framing is shown in Figure 1-19.

There are a number of similarities between T-1 and E-1 as well: Channels are all 64 Kbps, and frames are transmitted 8,000 times per second. Also, whereas T-1 gangs together 24 frames to create an extended superframe, E-1 gangs together 16 frames to create what is known as a *ETSI multiframe*. The multiframe is subdivided into two submultiframes; the CRC-4 in each one is used to check the integrity of the submultiframe that preceded it.

A final word about T-1 and E-1: Because T-1 is a departure from the international E-1 standard, it is incumbent upon the T-1 provider to perform all interconnection conversions between T-1 and E-1 systems. For example, if a call arrives in the United States from a European country, the receiving American carrier must convert the incoming E-1 signal to T-1. If a call originates from Canada and is terminated in Australia, the Canadian originating carrier must convert the call to E-1 before transmitting it to Australia.

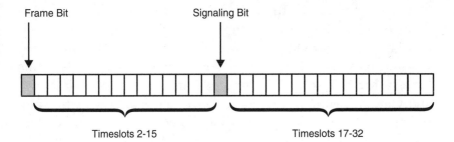

FIGURE 1-19   E-1 framing.

CABLE-BASED ACCESS TECHNOLOGIES

In 1950, Ed Parsons placed an antenna on a hillside above his home in Washington State, attached it to a coaxial cable distribution network, and began to offer television service to his friends and neighbors. Prior to his efforts, the residents of his town were unable to pick up broadcast channels because of the blocking effects of the surrounding mountains. Thanks to Parsons, *community antenna television* (CATV) was born; from its roots came cable television.

Since that time, the cable industry has grown into a $43 billion industry. In the United States alone, 10,000 headends deliver content to 70 million homes in more than 25,000 communities over more than a million miles of coaxial and fiber-optic cable. As the industry's network has grown, so too have the aspirations of those deploying it. Their goal is to make it much more than a one-way medium for the delivery of television and pay-per-view; they want to provide a broad spectrum of interactive, two-way services that will enable them to compete head-to-head with the telephony industry. To a large degree, they are succeeding. The challenges they face, however, are daunting.

THE BROADBAND BALKANS

Unlike the telephone industry that began under the control and design direction of a small number of like-minded individuals (Alexander Bell and Theodore Vail, among others), the cable industry came about thanks to the combined efforts of hundreds of innovators, each of them building on Ed Parson's original idea. As a consequence, the industry, although enormous, was in many ways fragmented. Powerful industry leaders like Liberty Media's John Malone and Time Warner's Ted Turner and Gerald Levine were able to exert Tito-like powers to unite the many companies, turning a loosely cobbled-together collection of players into cohesive, powerful corporations with a shared vision of what they were capable of accomplishing.

Today, the cable industry is a force to be reckoned with, and upgrades to the original network are well underway. This is a crucial activity that will ensure the success of the industry's ambitious business plan and provide a competitive balance for the traditional telcos. Cable companies are living up to their promise of becoming broadband data providers—they have deployed digital cable to more than 14 million subscribers, 7 million of whom also subscribe to cable modem service for Internet access. Today, cable modems are winning the race against DSL, although how long that will continue is uncertain.

THE CABLE NETWORK

The traditional cable network is an analog system based on a tree-like architecture. The headend, which serves as the signal origination point, is connected to the downstream distribution network by a 1-inch diameter rigid coaxial cable, as shown in Figure 1-20. That cable delivers the signal, usually a 450-MHz collection of 6 MHz channels, to a neighborhood where splitters divide the signal and send it down half-inch diameter semi-rigid coax that typically runs down a residential street. At each house, another splitter pulls off the signal and feeds it to the

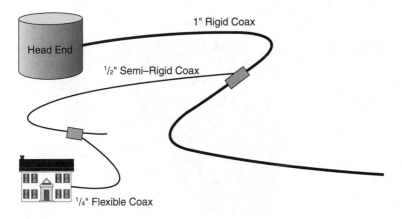

FIGURE 1-20   Layout of typical cable distribution network.

set-top box in the house over the drop wire, a local loop of flexible quarter-inch coaxial cable. A typical splitter is shown in Figure 1-21.

Although this architecture is perfectly adequate for the delivery of one-way television signals, its shortcomings for other services should be fairly obvious to the reader. First of all, it is, by design, a broadcast system. It does not typically have the capability to support upstream traffic (from the customer toward the headend) and is therefore not suited for interactive applications. Second, because of its design, the network is prone to significant failures that have the potential to affect large numbers of customers. The tree structure, for example, means that if a failure occurs along any branch in the tree, every customer from that point downward loses service. Contrast this with the telephone network where customers have a dedicated local loop over which their service is delivered. Second, because the system is analog, it relies on amplifiers to keep the signal strong as it is propagated downstream. These amplifiers are powered locally—they do not have access to central office power as devices in the telephone network do. Consequently, a local power failure can bring down the network's capability to distribute service in that area.

FIGURE 1-21   Signal splitter in residential cable installation.

The third issue is one of customer perception. For any number of reasons, some of them technically valid, there is a general belief that the traditional cable network is not as capable or as reliable as the telephone network. As a consequence of this perception, the cable industry is faced with the daunting challenge of convincing potential voice and data customers that they are in fact capable of delivering high-quality service. Some of the concerns are justified. In the first place, the telephone network has been in existence for almost 125 years, during which time its operators have learned how to optimally design, manage, and operate it in order to provide the best possible service. The cable industry, on the other hand, came about 50 years ago and didn't benefit from the rigorously administered, centralized management philosophy that characterized the telephone industry. Additionally, typical 450-MHz cable systems did not have adequate bandwidth to support the bidirectional transport requirements of new services.

Furthermore, the architecture of the legacy cable network, with its distributed power delivery and tree-like distribution design, does not lend itself to the same high degree of redundancy and survivability that the telephone network offers. Consequently, cable providers have been hard-pressed to convert customers who have become vigorously protective of their telecommunications services.

NEXT-GENERATION CABLE SYSTEMS

Faced with these harsh realities and the realization that the existing cable plant could not compete with the telephone network in its original analog incarnation, cable engineers began a major rework of the network in the early 1990s. Beginning with the headend and working their way outward, they progressively redesigned the network to the extent that in many areas of the country, their coaxial local loop is capable of competing on equal footing with the telco's twisted pair—and in some cases besting it.

The process they have used to reach this rather remarkable stage in their competitive technological development comprises four phases. In the first phase, they converted the content generation process, that is, the headend, from analog to digital. This enabled them to digitally compress the content, resulting in far more efficient utilization of the available system bandwidth. Second, they undertook an ambitious physical upgrade of the coaxial plant, replacing the 1-inch trunk and half-inch distribution cable with optical fiber. This brought about several desirable results. First, by using a fiber feeder infrastructure, network designers were able to eliminate a significant percentage of the amplifiers responsible for such a high percentage of the failures the network experienced due to power problems in the field. Second, the fiber makes it possible to provision significantly more bandwidth than coaxial systems enable. Third, because the system is now largely digital, it suffers less from noise-related errors than its analog predecessor does. Finally, an upstream return channel, illustrated in Figure 1-22, was provisioned, which makes possible the delivery of true interactive services such as voice, Web surfing, and videoconferencing.

The third phase of the conversion had to do with the equipment provisioned at the user's premises. The analog set-top box has largely been replaced with a digital device that has the capability to take advantage of the digital nature of the network, including access to the upstream channel. It decompresses digital content, performs content (stuff) separation, and provides the network interface point for data and voice devices. Between

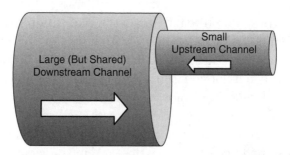

FIGURE 1-22   Upstream and downstream channels in modern cable systems.

the digital set-top box and the bandwidth made possible by fiber, cable systems routinely offer as many as 200 television channels, dozens of pay-per-view channels, and hundreds of music channels over their networks.

The last phase of the overall process is business conversion. Cable providers look forward to the day when their networks will compete on equal footing with the twisted pair networks of the telephone company, and customers will see them as viable competitors. In order for this to happen, they must demonstrate that their network is capable of delivering a broad variety of competitive services, that the network is robust, that they have *operations support systems* (OSSs) that will guarantee the robustness of the infrastructure, and that they are cost competitive with incumbent providers. All of the major providers have come to grips with this and are designing OSS infrastructures that will enable them to deliver the highest possible customer service quality over their networks.

HYBRID FIBER COAX SYSTEMS (HFC)

The current breed of cable delivery networks that shows the most promise is called a *Hybrid Fiber Coax* (HFC) system. As the name implies, HFC systems comprise a fiber feeder architecture that delivers broadband services to a served area. In the neighborhood or at the business location to which the service is being delivered, the fiber terminates at a hub. There, the optical signal is converted to electrical and delivered to the premises on traditional coaxial cable.

This architecture has several advantages. First of all, the optical feeder can be provisioned on a dual fiber ring architecture, optimally running the *Synchronous Optical Network* (SONET) physical layer protocol. In most fiber ring deployments, one ring is designated as the primary path, while the other is provisioned as a backup. SONET provides a feature known as *Automatic Protection Switching* (APS), which constantly monitors the integrity of the rings over which it is deployed. If the primary ring fails for any reason (backhoe fade[9] being the most common), devices on the ring will automatically switch and send traffic to the backup ring, limiting the outage to approximately 50 ms.

Another advantage is cost. Signals transported across optical fiber degrade far more slowly and accumulate less noise than those transported on copper or coaxial cable, resulting in the ability to build far longer spans with fewer repeaters—the digital equivalent of an amplifier. Furthermore, the bandwidth available in these systems is quite high, and as a consequence, more services can be provisioned to a broader population of customers.

Most cable operators have designated different ranges of the available system bandwidth for the transport of both digital and analog services. The digital domain carries compressed video and audio, while the analog portion of the cable spectrum transports voice and data.

A word about cable telephony: In order to transport voice, cable systems must be interconnected at the headend to the PSTN. This means that local and long distance carriers must have *points of presence* (POPs) in cable provider headend locations so that calls can be handed off to the switched network. This process is already underway. In fact, most long-distance companies have entered into peering agreements with cable providers so that they can have a POP adjacent to the cable headend.

CABLE MODEMS

As cable providers have progressively upgraded their networks to include more fiber in the backbone, their offer to provide two-way access to the Internet was a natural next step. Enter the cable modem. This new technology offers downstream

---

[9]A euphemism used in the transport industry to describe what happens when a backhoe severs a cable while digging. The fade, of course, is instantaneous and usually catastrophic.

access speeds of up to 10 Mbps, and although availability is still spotty in some areas, the number is growing as cable providers respond to the growing demand for broadband access. Today, the cable network passes in front of approximately 100 million homes, a very respectable potential market for cable-based data services.

Cable modems provide an affordable option to achieve high-speed access to the Web, with current monthly subscription rates in the neighborhood of $40. They offer asymmetric access, that is, a much higher downstream speed than upstream, but for the majority of users, this does not represent a problem because the bulk of their use will be for Web surfing during which the bulk of the traffic travels in the downstream direction.

When a customer uses a cable modem to access the Web, a splitter at the service demarc point directs the signal to the cable modem, usually located adjacent to the user's PC. The cable modem, in turn, connects to an Ethernet card installed in the PC, and although all this can be accomplished by a reasonably knowledgeable subscriber, the work is usually done by a cable installer (the proverbial cable guy) for roughly $100 in installation fees. In most areas where the service is available, the cable modem is provided as part of the service package; the Ethernet card may or may not be included. If not, the card can be had for approximately $30.

Although cable modems do speed up access to the Web and other online services by hundreds of percentage points, there are a number of downsides that must be considered. First and foremost, cable modem service is targeted at the consumer market, not at the business user. As a result, service quality and repair intervals are not necessarily what most corporate customers expect from a service that their business depends upon. In fact, few cities can offer cable modem access to business customers in downtown areas. Second, business customers who have a preestablished relationship with an ISP and an already well-known and advertised e-mail address will have to change them because cable modem service is sold as part of an overall package, again targeted at the consumer market (in reality, the subscriber does not have to change; they would, however, have to now pay for two accounts). This reality is currently under attack by such online services as AOL who believe that as common carriers, the cable providers should be required to open their networks to all. The cable providers, in turn, argue that their own ISPs would be endangered if they were to allow such incursion. The courts are currently hearing arguments from both sides, and a decision will be forthcoming soon—stay tuned.

Another issue with cable modem service has to do with performance and is a function of the architecture of the cable system. In HFC architectures, the available bandwidth is shared among all subscribers attached to a particular node—hundreds in some cases. As a consequence, heavy usage by multiple subscribers could result in a serious slowdown in service, a phenomenon that is familiar to many cable subscribers today. Cable companies claim that they will overcome the problem through capacity buildouts; time will tell. Today there are significant service problems around the country because of the relative success of cable modem deployments. Cable companies have oversold their networks and are now faced with infrastructures that cannot handle the load. Overpromising and underdelivering does not go over well in an industry with customers already overly sensitive to service outages; they must take steps now to minimize the perception of poor service quality or they will create even more problems for themselves down the road.

The final concern that has been voiced about cable modems is one of security. Because the bandwidth is shared, it is conceivable that other subscribers could access the contents of other systems on that shared facility, and although this has not yet become a serious concern, it is valid. Cable modem subscribers should ensure that all utilities, such as print and file sharing, are disabled on their PCs to prevent this from happening. Furthermore, they should purchase and install a firewall software package to monitor intrusive attacks. As a point of reference, I have a cable modem at my home office; according to my firewall software, my system received 44 automated attacks (all repulsed) in the first hour it was online.

DATA OVER CABLE STANDARDS

As interest grew in the late 1990s for broadband access to data services over cable television networks, CableLabs®, working closely with the ITU and major hardware vendors, crafted a standard known as the *Data Over Cable Service Interface Specification* (DOCSIS). Designed to ensure interoperability among cable modems as well as to assuage concerns about data security over shared cable systems, DOCSIS has done a great deal to resolve marketplace issues.

Under DOCSIS, CableLabs® crafted a cable modem certification standard called *DOCSIS 1.0*, which guarantees that modems carrying the certification will interoperate with any headend equipment, are ready to be sold in the retail market, and will interoperate with other certified cable modems. Engineers from Askey, Broadcom, Cisco Systems, Ericsson, General Instrument, Motorola, Philips, 3Com, Panasonic, Digital Furnace, Thomson, Terayon, Toshiba, and Com21 participated in the development effort.

The DOCSIS 1.1 specification was released in April 1999 and included two additional functional descriptions that began to be implemented in 2000. The first specification details procedures for guaranteed bandwidth as well as a specification for QoS guarantees. The second specification is called *Baseline Privacy Interface Plus* (BPI+); it enhances the current security capability of the DOCSIS standards through the addition of digital certificate-based authentication and support for multicast services to customers. DOCSIS specifies downstream traffic data rates between 27 and 36 Mbps over a *radio frequency* (RF) path in the 50-MHz to 750+-MHz range and upstream data rates between 320 Kbps and 10 Mbps over a 5- to 42-MHz path. Because data over cable uses a shared loop, however, individuals will see rates decline as the number of users increases.

Although the DOCSIS name is in widespread use, CableLabs now refers to the overall effort as the CableLabs Certified Cable Modem Project.

DIGITAL SUBSCRIBER LINE (DSL)

The access technology that has enjoyed the greatest amount of attention recently is DSL. It provides an ideal solution for remote LAN access, Internet surfing, and access for telecommuters to corporate databases.

DSL came about largely as a direct result of the Internet's appearance in the public's consciousness in 1993. Prior to its arrival, the average telephone call lasted approximately four minutes, a number that central office personnel used while engineering switching systems to handle expected call volumes. This number was arrived at after nearly 125 years of experience designing networks for voice customers. They knew about Erlang theory, loading objectives, and peak-calling days/weeks/seasons, and had decades of trended data to help them anticipate load problems. So in 1993, when the Internet —and with it, the World Wide Web—arrived, network performance became unpredictable as callers began to surf, often for hours on end. The average four-minute call became a thing of the past as online service providers, such as AOL, began to offer flat-rate plans that did not penalize customers for long connect times. Then, the unthinkable happened: Switches in major metropolitan areas, faced with unpredictably high call volumes and hold times, began to block[10] during normal business hours, a phenomenon that only occurred in the past during disasters or on Mother's Day. Something had to be done.

A number of solutions were considered, including charging different rates for data calls than for voice calls, but none of these proved feasible until a technological solution was proposed. That solution was DSL.

It is a well-known fact among telephony personnel that the analog telephone network has been carefully designed around the known behavior patterns of the human voice. As we noted earlier (again, thanks to work done at Bell Laboratories), the

---

[10]Switches are designed in such a way that they will block the processing, incoming calls when they find themselves processing unusually high call volumes. This is done to ensure that emergency services can be maintained. While this is a process that happens by design, under normal circumstances, it rarely occurs.

human voice comprises a rich mixture of frequencies that range from approximately 20 Hz to 20 KHz. Although this broad collection of spectral components adds tremendous richness, tone, and timbre to the sound of the voice, the extremely high and low frequency components are not necessary for the human ear to recognize and understand what is being said by another person. In fact, a relatively narrow range of frequencies is required. This range, known as the *voice band*, spans the spectrum of frequencies that ranges between 300 and 3,300 Hz. As long as a telephone network can collect and deliver this portion of a caller's voice, the listener will be able to recognize and understand what the caller is saying.

This recognition was a real boon for the phone company because it meant that they could engineer their networks with far better frequency efficiency. Instead of building their systems with the intent of giving every subscriber a 20-KHz piece of spectrum, they were able to reduce that number to 4 KHz, thus enabling significantly more customers to be squeezed into whatever frequency range they had available. Thus, a 20-KHz piece of bandwidth could support approximately five customers instead of one.

It is commonly believed that because of the way the telephone network is designed, the local loop is incapable of carrying more than the frequencies required to support the voice band. This is patently untrue. ISDN, for example, requires significantly high bandwidth to support its digital makeup. When ISDN is deployed, the loop must occasionally be modified in certain ways to eliminate its designed bandwidth limitations. For example, some long loops are deployed with devices called *load coils*, which tune the loop to the voice band (they are effectively notch filters). They make the transmission of frequencies above the voice band impossible, but enable the relatively low-frequency voice band components to be carried across a long local loop. These load coils must be removed if digital services are to be deployed.

Thus, a local loop is only incapable of transporting high-frequency signal components if it is *designed* not to carry them. The capacity is still there; the network simply makes it unavailable. DSL services, especially ADSL, take advantage of this disguised bandwidth.

DSL TECHNOLOGY

In spite of the name, DSL is analog technology. The devices installed on each end of the circuit are sophisticated high-speed modems that rely on complex encoding schemes to achieve the high bit rates that DSL offers. Furthermore, several of the DSL services, specifically ADSL, G.Lite, the *Very High-Speed Digital Subscriber Line* (VDSL), and the *Rate-Adaptive Digital Subscriber Line* (RADSL), are designed to operate in conjunction with voice across the same local loop. ADSL is the most commonly deployed service and offers a great deal to both business and residence subscribers.

ADSL MODULATION TECHNIQUES

Two modulation techniques have emerged for the encoding of ADSL signals. These are *Discrete Multitone Modulation* (DMT) and *Carrier-Suppressed Amplitude Phase Modulation* (CAP). ISDN BRI's *2 Binary 1 Quaternary* (2B1Q) line code was an early entrant in the DSL world as the standard for *ISDN DSL* (IDSL), but it has become more of a niche scheme today compared to the other two. It encodes two bits per transmitted signal and thus can achieve bit rates that are twice the signaling rate. (Refer to the discussion on bit rate versus baud earlier in this chapter.)

CROSSTALK ISSUES

Because many DSL services use the standard local loop wire pair, there must be some technique in place to prevent crosstalk (channel bleeding) between the upstream and downstream sides of the circuit. Two techniques are commonly used: FDM, in which separate frequency ranges are established for the two transmitted components, or FDM combined with echo cancellation. In FDM systems, voice is relegated to the 0- to 4-KHz band, upstream traffic to a 25- to 200-KHz band, and downstream traffic to a 200-KHz to 1.1-MHz band. Thus, the three components are completely isolated from one another. In echo-cancelled systems, the voice traffic remains in the 0- to 4-KHz band, but the upstream and downstream traffic share access to the 25-KHz to 1.1-MHz band, hence the need for echo cancellation.

Echo cancellation is required whenever both ends of a circuit share access to the same range of available bandwidth. When one end transmits to the other end, it is common for some of the transmitted signal to be reflected back to the transmitter because of impedance mismatches. When this happens, the reflected signal can be mistaken for information received from the far end. To combat this, receivers remove the signal transmitted from anything they receive to ensure that what they receive is a pure transmitted signal from the far end.

Generally speaking, CAP-based systems use FDM, while DMT-based systems combine FDM with echo cancellation.

DMT

DMT is a relatively new encoding scheme that divides the available bandwidth into 256 4-KHz subcarriers, as shown in Figure 1-23. The information to be transmitted is then encoded into these subchannels based upon the relative performance of each channel. In other words, before information is encoded into the subcarriers, they are tested for relative performance based on noise, and so on. Those with high noise coefficients are not utilized. Obviously, the more clean channels that are available, the higher the aggregate bandwidth. This results in an extremely granular assignment of bandwidth and high-transmission quality.

DMT has been designated by the *American National Standards Institute* (ANSI) as the standard for ADSL (ANSI T1.413-1995) and as such enjoys significant popularity throughout the industry as an open standard.

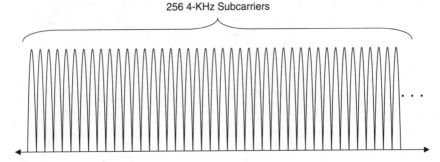

256 4-KHz Subcarriers

FIGURE 1-23   Discrete Multitone Transmission (DMT).

CAP

CAP is a derivative of a well-known encoding scheme called *quadrature amplitude modulation*, or QAM (pronounced kwam). It uses a combination of amplitude and phase modulation to achieve high bits-persignal encoding levels, an example of which is shown in Figure 1-24. CAP has the advantages of being more mature in the marketplace than DMT, significantly simpler than DMT, and better at echo cancellation, but it suffers from the serious disadvantage of being available from only a single source —GlobeSpan Semiconductor Corporation. Nevertheless, there is a significant amount of support for CAP in the marketplace in spite of the fact that it is a proprietary solution.

Enough about technology; on to services.

DSL SERVICES

DSL comes in a variety of flavors, all designed to provide flexible, efficient, high-speed service across the existing telephony infrastructure. From a consumer point of view, DSL, especially ADSL, offers a remarkable leap forward in terms of the available bandwidth for broadband access to the Web. As content has steadily moved away from being largely text-based and has become more image-based, the demand for faster delivery services is a cry that has been growing in intensity for some time

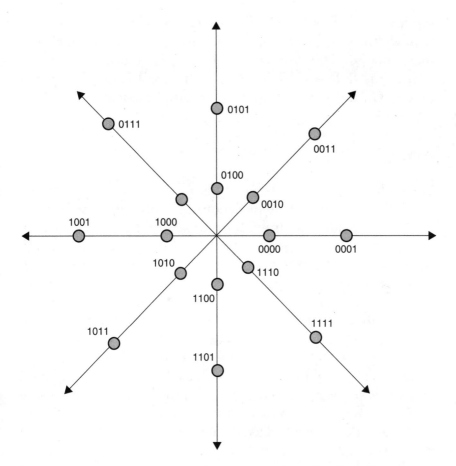

FIGURE 1-24   Carrierless Amplitude Phase Modulation (CAP).

now. DSL may provide the solution at a reasonable cost to both the service provider and the consumer.

Businesses, on the other hand, also stand to benefit from DSL technology rollouts. Remote workers and telecommuters can rely on DSL for remote LAN and Internet access as well as for specialized services. A medical specialist can use DSL to view digitized medical images, a remote architect to view complex CAD files, or a film editor to review a digitized film clip. Furthermore, DSL provides a good technology solution for VPN access as well as for ISPs looking to grow the bandwidth available to their customers. DSL is available in a variety of both symmetric and asymmetric services, and thus offers a high-bandwidth access solution for a variety of applications. The most common DSL services are ADSL, *high bit rate digital subscriber line* (HDSL), HDSL-2, RADSL, and VDSL. The special case of G.Lite, a form of ADSL, will also be discussed.

ADSL

When the World Wide Web and flat rate access charges arrived in all their splendor, the typical consumer phone call went from roughly four minutes in duration to several times that. All the engineering that had led to the overall design of the network based on an average four-minute hold time went out the window as the switches staggered under the added load. Never was the expression, "In its success lie the seeds of its own destruction," more true. When ADSL arrived, it provided the offload that was required to save the network.

The typical ADSL installation is shown in Figure 1-25. No change is required to the two-wire local loop; minor equipment changes, however, are required. First, the customer must have an ADSL modem at their premises. This device enables both the telephone service and a data access device, such as a PC, to be connected to the line.

The ADSL modem is more than a simple modem in that it also provides the FDM process required to separate the voice and data traffic for transport across the loop.[11] When voice traffic reaches the ADSL modem, it is immediately encoded in the traditional voice band frequencies that will be handed off to the local switch upon arrival at the central office. The modem is often referred to as an *ADSL transmission unit for remote use*

---

[11]The device that actually does this is called a *splitter* in that it splits the voice traffic away from the data. It is usually bundled as part of the ADSL modem, although it can also be installed as a card in the PC, as a standalone device at the demarc point, or as a series of distributed devices on each phone at the premises. The most common implementation today is to have the splitter integrated as part of the DSL modem; this, however, is the least desirable implementation because this design can lead to crosstalk between the voice and data circuitry inside the device.

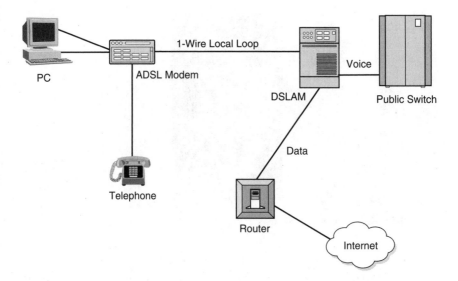

FIGURE 1-25   ADSL circuit layout.

(ATU-R). Similarly, the device in the central office is often called an *ATU-C* (for central office).

When a PC wants to transmit data across the local loop for Internet access or some other application, the traffic is encoded in the higher-frequency band that is reserved for data traffic. The ADSL modem knows to do this because the traffic is arriving on a port that is reserved for data devices. Upon arrival at the central office, the data traffic does not travel to the local switch; instead, it stops at the ADSL modem that has been installed at the central office end of the circuit. In this case, the modem serves a large number of subscribers and is known as a DSLAM (pronounced dee-slam). A bank of DSLAMs is shown in Figure 1-26.

This is where the magic of DSL really shines. Instead of traveling on to the local switch, the data traffic is passed around the switch, a process known as a *line-side redirect*. The traffic is passed to a router, which in turn is connected to the Internet.

The advantages of this architecture are legion. First, it offloads the data traffic from the local switch, which can then

FIGURE 1-26   A bay of DSLAMs.

go back to doing what it does best—switching voice traffic. Second, it creates a new line of business for the service provider. As a result of adding the router and connecting the router to the Internet, the service provider instantly becomes an ISP. In keeping with the spirit of convergence, this is a near-ideal combination because it enables the service provider to become a true service provider by offering much more than simple access and transport.

As the name implies, ADSL provides two-wire asymmetric service—that is, the upstream bandwidth is different from the downstream. In the upstream direction, data rates vary from 16 to 640 Kbps, while the downstream bandwidth varies from 1.5 to 8 Mbps. Because most applications today are asymmetric in nature, this disparity poses no problem for the average consumer of the service.

A WORD ABOUT THE DSLAM

This device has received a significant amount of attention recently because of the central role that it plays in the deployment of broadband access services. Obviously, the DSLAM must interface with the local switch so that it can pass voice calls on to the PSTN. However, it often interfaces with a number of other devices as well. For example, on the customer side, the DSLAM may connect to a standard ATU-C, directly to a PC with a built-in *network interface card* (NIC), to a variety of DSL services, or to an integrated access device of some kind. On the trunk side (facing the switch), the DSLAM may connect to IP routers as described before, to an ATM switch, or to some other broadband service provider. It, therefore, becomes the focal point for the provisioning of a wide variety of access methods and service types.

HIGH BIT RATE DIGITAL SUBSCRIBER LINE (HDSL)

The greatest promise of HDSL is that it provides a mechanism for the deployment of four-wire T1 and E1 circuits without the need for span repeaters, which can add significantly to the cost of deploying data services. It also means that service can be deployed in a matter of days rather than weeks, something customers certainly applaud.

DSL technologies in general enable for repeaterless runs of wire as far as 12,000 feet, while traditional four-wire data circuits, such as T-1 and E-1, require repeaters every 6,000 feet. Consequently, many telephone companies are now deploying HDSL behind the scenes as a way to deploy these traditional services. Customers do not realize that the T-1 facility they are plugging their equipment into is being delivered using HDSL technology. The important thing is that they don't *need* to know. All the customer should have to care about is that there is now a SmartJack installed in the basement, and through that jack they have access to 1.544 Mbps or 2.048 Mbps of bandwidth—period.

HDSL2

HDSL2 offers the same service that HDSL offers, with one added (and significant) advantage: It does so over a single pair of wire rather than two. It also provides other advantages. First, it was designed to improve vendor interoperability by requiring less equipment at either end of the span (transceivers and repeaters). Second, it was designed to work within the confines of standard telephone company *Carrier Serving Area* (CSA) guidelines by offering a 12,000-foot wire-run capability that matches the requirements of CSA deployment strategies. (See the discussion on CSA guidelines later in this section.)

A number of companies, including *Competitive Local Exchange Carrier* (CLEC) Rhythms NetConnections, have deployed T-1 access over HDSL2 at rates 40 percent lower than typical T-Carrier prices. Allegiance Telecom is working with equipment manufacturer Adtran to determine the range of services that HDSL2 will provide when deployed. Furthermore, a number of vendors including Lucent, 3Com, Nortel Networks, and Alcatel have announced their intent to work together to achieve interoperability among DSL modems by working with the University of New Hampshire interoperability test labs.

RADSL

RADSL (pronounced Rad-zel) is a variation of ADSL designed to accommodate changing line conditions that can affect the overall performance of the circuit. Like ADSL, it relies on DMT encoding, which selectively populates subcarriers with transported data, thus enabling for granular rate-setting.

VDSL

VDSL shows promise as a provider of extremely high levels of access bandwidth—as much as 52 Mbps over a short local loop. VDSL requires *fiber-to-the-curb* (FTTC) architecture and recommends ATM as a switching protocol; from a fiber hub, copper tail circuits deliver the signal to the business or residential premises. Bandwidth available through VDSL ranges from 1.5 to 6 Mbps on the upstream side and from 13 to 52 Mbps on the downstream side. Obviously, the service is distance sensitive and actual achievable bandwidth drops as a function of distance. Nevertheless, even a short loop is respectable when such high-bandwidth levels can be achieved. With VDSL, 52 Mbps can be reached over a loop length of up to 1,000 feet—not an unreasonable distance by any means.

G.LITE

Because the installation of splitters has proven to be a contentious and problematic issue, the need arose for a version of ADSL that did not require them. That version is known as either *ADSL Lite* or *G.Lite* (after the ITU-T G-Series standards[12] that govern much of the ADSL technology). In 1997, Microsoft, Compaq, and Intel created the *Universal ADSL Working Group* (UAWG),[13] an organization that grew to nearly 50 members dedicated to the task of simplifying the rollout of ADSL. In effect, the organization had four stated goals:

To ensure that analog telephone service will work over the G.Lite deployment without remote splitters in spite of the fact that the quality of the voice may suffer slightly due to the potential for impedance mismatch.

To maximize the length of deployed local loops by limiting the maximum bandwidth provided. Research indicates that customers are far more likely to notice a performance improvement when migrating from 64 Kbps to 1.5 Mbps than when going from 1.5 Mbps to higher speeds. Perception is clearly important in the marketplace, so the UADSL Working Group chose 1.5 Mbps as their downstream speed.

To simplify the installation and use of ADSL technology by making the process as plug-and-play as possible.

To reduce the cost of the service to a perceived reasonable level.

[12]ITU-T stands for International Telecommunications Union-Telecommunications Standardization Sector and is the organization that replaced the CCITT in 1992.

[13]The group self-dissolved in the summer of 1999 after completing what they believed their charter to be.

Most DSL vendors are working on ways to achieve success with DSL in the marketplace, and G.Lite is high on various lists as a winning option. At the 1999 annual DSL Conference and Exhibition in Reston, Virginia, Steve Rago of Lucent Technologies identified six key challenges that must be overcome if DSL is to be a viable contender for broadband access. They are

- Service provisioning
- Reliability
- Suitability of loop plant
- Subscriber scalability
- Power management
- Performance

Ultimately, we will see the first two challenges addressed through the use of integrated voice and data line cards. The second two will be overcome through the deployment of both G.Lite and full-service ADSL, depending on the nature of the loop; the final two will be achieved through the deployment of superior microelectronics.

Of course, G.Lite is not without its detractors. A number of vendors have pointed out that if G.Lite requires the installation of microfilters at the premises on a regular basis, then true splitterless DSL is a myth because microfilters are in effect a form of splitter. They contend that if the filters are required anyway, then they might as well be used in full-service ADSL deployments to guarantee high-quality service delivery. Unfortunately, this flies in the face of one of the key tenets of G.Lite, which is to simplify and reduce the cost of DSL deployment by eliminating the need for an installation dispatch (a truck roll in the industry's parlance). The key to G.Lite's success in the eyes of the implementers is to eliminate the dispatch, minimize the impact on traditional POTS telephones, reduce costs, and extend the achievable drop length. Unfortunately, customers still have to be burdened with the installation of microfilters, and coupled noise on POTS is higher than expected. Many vendors argue that these problems largely disappear with full-feature ADSL using splitters; a truck dispatch is still required, but again, it is often required to install the microfilters anyway, so there is no net loss. Furthermore, a number of major semiconductor manufacturers support both G.Lite and ADSL on the same chipset, so the decision to migrate from one to the other is a simple one that does not necessarily involve a major replacement of internal electronics.

SHDSL

A new flavor of DSL is just hitting the marketplace in the form of the *Symmetric High-Density Digital Subscriber Line* (SHDSL). This standard is based on the preexisting standards that governed SDSL, HDSL, HDSL-2, IDSL, and others, and therefore is easily implementable and well understood. It offers 30 percent longer reach than earlier versions of DSL, a fact that overcomes the single greatest barrier to DSL's widespread deployment. It also offers symmetric bandwidth over a two-wire local loop at rates in excess of 2 Mbps, while a four-wire version can offer bandwidth in the 5-Mbps range. The technology is rate-adaptive and can, therefore, afford service providers the ability to offer a wide array of service plans. The primary market for SHDSL is the *small-to-medium* (SMB) business sector; it supports multiservice deployments including voice, video, data, and videoconferencing.

DSL MARKET ISSUES

DSL technology offers advantages to both the service provider and the customer. The service provider benefits from successful DSL deployment because it serves not only as a cost-effective technique for satisfying the bandwidth demands of customers in a timely fashion, but also because it provides a Trojan horse approach to the delivery of certain preexisting services. For example, many providers today implement T-1 and E-1 services over HDSL because it is cost effective. Although HDSL-2 is not yet widely deployed, it soon will be, at which time the advantages for the service provider will be even more enhanced as a result of the two-wire rather than four-wire nature of the service. Customers are blissfully unaware of the fact; in this case, it is the service provider rather than the customer who benefits most from the deployment of the technology. Of course, the accelerated installation interval is very much to the customer's advantage. This phenomenon lies at the heart of convergence: there is nothing wrong with service providers enjoying benefits from converged technologies because it facilitates their ability to provide advantages to their own customers.

From a customer point of view, DSL provides a cost-effective way to buy medium-to-high levels of bandwidth and, in some cases, embedded access to content. In 1999, AOL announced strategic partnerships with both Bell Atlantic and SBC Corporation under which customers could buy DSL access to AOL, bundled for $40 per month. This proved to be a boon for both Bell Atlantic and SBC because it gave them the coveted relationship with a content provider that they desired. It also gave them a ready market for their DSL services, a market that they had been somewhat reluctant to enter because of the current regulatory environment. Under the terms of the tariffs that guide their activities in their operating regions, the *Incumbent Local Exchange Carriers* (ILECs, the former *Regional Bell Operating Companies* [RBOCs]) must sell services like DSL at a discount to their wholesale customers, many of whom are their direct in-region competitors. The danger lies in the fact that those customer/competitors could easily buy unbundled DSL facilities from the incumbent company, form a long-distance relationship with a bandwidth provider like Qwest, and take the ILEC completely out of the service equation. This is clearly an undesirable outcome for the ILECs, so the opportunity to partner with the likes of AOL offers a somewhat protected point of penetration for DSL services.

Furthermore, the ILECs enjoy a substantial revenue stream from T-1 services that operate at 1.544 Mbps. If DSL service can provide the same bandwidth at a substantially lower cost to the customer, then T-1 revenues could be endangered by the widespread deployment of DSL. This concern is clearly in the forefront of the minds of the local incumbent service providers, but they also recognize that in this chaotic market, it is sometimes necessary to cannibalize some traditional market segments in order to maximize long-term profit in others.

Other forces have affected the DSL marketplace as well. In November 1999, the FCC ruled that local telephone companies must unbundle their local loops to the extent that competitors can share access to the copper so as to provide DSL services, even if the local carrier provides the voice service over the same line. In other words, under the terms of the ruling, a customer in Dallas, for example, could theoretically buy his voice service from SBC and high-speed data access service (DSL) from Covad Communications Group, both delivered across the customer's existing local loop. The Commission's intent with the line sharing ruling is to increase the number of companies offering high-speed data services while at the same time accelerating and geographically broadening the deployment of DSL. ILECs already implement line sharing with their own voice and DSL offerings, and because DSL is a form of FDM, it lends itself to this kind of sharing arrangement among competitors that share access to a common physical infrastructure. Customers benefit from this arrangement because they are not required to order a second line for DSL if they choose to purchase the service from a competitive provider, whereas they were prior to the ruling. ILECs also benefit because their own DSL services are still in the running as competitive offerings via the local loop. This model, however, has largely failed in the marketplace. Companies like Covad, Northpoint, and Rhythms NetConnections have essentially failed because of a single, clearly illuminated fact: Because of the current regulatory environment, local service providers *must* own the entire network if they are to succeed. Because they do not, the companies listed previously have no hope to compete with the ILECs. This will be discussed later when we cover telecomm regulation.

DSL also represents the primary bastion of defense against the ongoing market penetration efforts of cable modems. Currently, cable modems are taking the broadband access market by storm because the technology is widely available and works reasonably well—noticeably better (read faster) than traditional analog modem technology. DSL has yet to achieve the penetration levels that cable modems enjoy; a number of technical issues, including loop length and age of the outside plant, are having a significant negative impact on widespread deployment. Cable modems are expected to lead the broadband access charge for the next few years, but DSL is expected to catch up and eventually bypass the level of cable penetration. In fact, the ILECs have for the most part made aggressive plans to roll out DSL on a large scale.

Another contributing factor in favor of the ILECs is ubiquity. The cable industry's network passes within the service distance of more than 95 percent of the homes and businesses in the country. However, their actual market penetration is substantially lower than that. The telephone company's twisted pair network on the other hand is very close to universally deployed, with substantially higher customer penetration levels. Furthermore, many ILECs have crafted service alliances with direct broadcast television satellite providers, and between the soon-to-be-available two-way data transmission that these satellites will provide and the in-place high-speed service that DSL makes possible, the incumbents will most likely remain the provider of choice for both voice and data services. According to a recent Price-Waterhouse study on who customers would select as their residential service provider, the ILECs win the game:

ILEC	40%
IEC	21%
Satellite	7%
CATV	7%
Power	5%
Wireless	2%

Business customers, on the other hand, expressed very different numbers in the study: 36 percent of them indicated that they would switch to an *interexchange carrier* (IEC) or CLEC if given the opportunity, based purely on service—not technology. This is a challenge that the ILECs must remain cognizant of as they go forward with their competitive positioning plans.

PROVIDER CHALLENGES

The greatest challenges facing those companies looking to deploy DSL are competition from cable, unfettered pent-up customer demand, installation issues, and plant quality.

Competition from cable companies represents a threat to wireline service providers on several fronts. First, cable modems enjoy a significant amount of press and are therefore gaining well-deserved marketshare. The service they provide, although spottily available, is well received in the areas served and offers high-quality, high-speed access.

The second challenge is unmet customer demand. If DSL is to satisfy the broadband access requirements of the marketplace, it must be made available throughout ILEC service areas. This means that incumbent providers must equip their central offices with DSLAMs that will provide the line-side redirect required as stage one of DSL deployment. Again, the law of primacy rears its head here; the ILECs must get to market quickly with their own broadband offerings if they are to attain and retain a toehold in the burgeoning broadband access marketplace.

One thing that will help them achieve primacy is the strategic relationship some ILECs have made with service providers. Bell Atlantic, GTE, and SBC have forged alliances with AOL, mentioned earlier, to provide DSL-based access to the service, bundled for about $40 per month. This form of technology and service convergence represents an advantage for everyone: Customers benefit from having one-stop shopping for both access and service, AOL gets enhanced access to ILEC customers, and the ILECs find a safe harbor within which to deploy DSL. Furthermore, AOL is aligned with DirecTV and DirecPC, both of which provide alternative wireless access methods. If the ILECs become part of that alliance, particularly in the face of cable's refusal to enable the AOLs of the world ready access to their networks, the ILECs and their partners could achieve marketplace primacy over the broadband cable providers. Again, time will tell. Manufacturers stepped up to the challenge of rapid deployment with central office-based devices that support SDSL, ADSL, G.Lite, and HDSL2. They offer a high-speed ATM switching fabric that enables service providers to offer diverse and granular levels of QoS, something the market is keen about. They also offer support for both frame relay and ATM access and often incorporate a Wizard-based configuration and NM system.

The third and fourth challenges to rapid and ubiquitous DSL deployment are installation issues and plant quality. A significant number of impairments have proven to be rather vexing for would-be deployers of widespread DSL. These challenges fall into two categories: electrical disturbances and physical impairments.

ELECTRICAL DISTURBANCES

The primary cause of electrical disturbance in DSL is crosstalk, caused when the electrical energy carried on one pair of wires bleeds over to another pair and causes interference (noise) there. Crosstalk exists in several flavors. *Near-end crosstalk* (NEXT) occurs when the transmitter at one end of the link interferes with the signal received by the receiver at the same end of the link, while *far-end crosstalk* (FEXT) occurs when the transmitter at one end of the circuit causes problems for the signal received by a receiver at the far end of the circuit. Similarly, problems can occur when multiple DSL services of the same type exist in the same cable and interfere with one another. This is referred to as *self-NEXT* or *self-FEXT*. When different flavors of DSL interfere with one another, the phenomenon is called *foreign-NEXT* or *foreign-FEXT*. Other problems that can cause errors in DSL and therefore a limitation in the maximum achievable bandwidth of the system are simple *radio frequency interference* (RFI), which can find its way into the system, as well as impulse, Gaussian, and random noise that exist in the background but can affect signal quality even at extremely low levels.

PHYSICAL IMPAIRMENTS

The physical impairments that can have an impact on the performance of a newly deployed DSL circuit tend to be characteristics of the voice network that typically have a minimal effect on simple voice and low-speed data services. These include load coils, bridged taps, splices, mixed gauge loops, and weather conditions.

LOAD COILS   These present a problem for DSL deployment because of the nature of their internal anatomy. Load coils are nothing more than lumped inductance installed on long loops to overcome distance-related impairments to voice. In effect, load coils are low-pass filters that tune the local loop to the voice band frequency range. Because the high-frequency components of the human voice are not required, the load coils filter them out, leaving nothing but the lower-frequency voice band. Also, because low-frequency signals propagate farther than high-frequency signals,[14] the addition of load coils enables the frequencies required to understand and recognize the received voice to be carried the maximum distance possible without worrying about the smearing that would occur if the higher-frequency components were included in the mix.

Unfortunately, load coils effectively eliminate the high-frequency transmission channel required to transport the broadband data segment of the DSL service. They must,

---

[14]Think about the last time that you attended a parade. You may recall that the event that notified you of the imminent arrival of the parade, led by the band, was the low-frequency sound of the bass drums—not the brass section. The drum sounds reached your ears first.

therefore, either be removed prior to deploying DSL across a loaded loop, or another loop that is not loaded must be used.

BRIDGED TAPS   When a multipair telephone cable is installed in a newly built-up area, it is generally some time before the assignment of each pair to a home or business is actually made. To simplify the process of installation when the time comes, the cable's pairs are periodically terminated in terminal boxes (sometimes called *B-Boxes*) installed every block or so. There may be multiple appearances of each pair on a city street, waiting for assignment to a customer. When the time comes to assign a particular pair, the installation technician will simply go to the terminal box where that customer's drop appears and cross-connect the loop to the appearance of the cable pair (a set of lugs) to which that customer has been assigned. This eliminates the need for the time-consuming process of splicing the customer directly into the actual cable.

Unfortunately, this efficiency process also creates a problem. Although the customer's service has been installed in record time, there are now unterminated appearances of the customer's cable pair in each of the terminal boxes along the street. Although these so-called bridged taps present no problem for analog voice, they can be a catastrophic source of noise due to signal reflections that occur at the copper-air interface of each bridged tap. If DSL is to be deployed over the loop, the bridged taps must be removed, and although the specifications indicate that bridged taps do not cause problems for DSL, actual deployment says otherwise. To achieve the bandwidth that DSL promises, the taps must be terminated.

SPLICES AND GAUGE CHANGES   Although they cause less of a problem than some of the other impairments, splices can result in service impairments due to cold solder joints, corrosion, or weakening effects caused by the repeated bending of wind-driven aerial cable.

Gauge changes tend to have the same effects that plague circuits with unterminated bridged taps. When a signal traveling down a wire of one gauge jumps to a piece of wire of another gauge, the signal is reflected, resulting in an impairment known as *intersymbol interference*. The use of multiple gauges is common in a loop deployment strategy because it enables outside plant engineers to use lower-cost, small-gauge wire where appropriate, cross-connecting it to larger-gauge, lower-resistance wire where necessary.

WEATHER CONDITIONS   This is perhaps a bit of a misnomer. The real issue is moisture. One of the greatest challenges facing the deployment of DSL is the age of the loop plant. Much of the older distribution cable uses inadequate insulation between the conductors (paper in some cases). In some cases the outer sheath has cracked, enabling moisture to seep into the cable itself. The moisture causes crosstalk between wire pairs that can last until the water evaporates; unfortunately, this can take a considerable amount of time and result in extended outages.

SOLUTIONS

All of these factors have solutions. The real question is whether they can be remedied at a cost that is within reasonably achievable bounds. Given the growing demand for broadband access, there seems to be little doubt that the elimination of these factors would be worthwhile at all reasonable costs, particularly considering how competitive the market for the local loop customer has become.

The electrical effects, largely caused by various forms of crosstalk, can be reduced or eliminated in a variety of ways. Physical cable deployment standards are already in place that, when followed, help to control the amount of near and far-end crosstalk that can occur within binder groups in a given cable. Furthermore, filters have been designed that eliminate the background noise that can creep into a DSL circuit.

Physical impairments can be controlled to a point, although to a certain extent the service provider is at the mercy of their installed network plant. Obviously, older cable will present more physical impairments than newer cable, but there are steps that the service provider can take to maximize their success rate when entering the DSL marketplace. The first step that they can take is to prequalify local loops for DSL service to the greatest extent possible. This means running a series of tests using *mechanized loop testing* (MLT) to determine whether each loop's transmission performance falls within the bounds established by the service provider, existing standards, and industry-support organizations.

For DSL prequalification, MLT tests the following performance indicators (and others as required):

· Cable architecture
· Loop length
· Crosstalk and background noise

PREVENTATIVE NETWORK ARCHITECTURE ISSUES

New cable architecture is related to the actual deployment plan of the loop being tested. Until the mid-1980s, a criterion known as the *Resistance Design Rule* dictated that the maximum loop resistance enabled would be 1500 ohms, with no limit on the number of bridged taps or gauge changes. This made sense, considering that the plant was intended and designed for the transport of voice and low-speed and voice-band data.

After the mid-1980s, things changed. Instead of focusing exclusively on resistance as the primary loop design criterion, outside plant engineers shifted to the use of CSA Guidelines that limit both the number of gauge changes and the maximum loop length, as well as the maximum length of any bridged taps that are present. This design came in concert with the widespread deployment of digital loop carrier architectures, which came about as a way to reduce the cost of deploying service to far-flung parts of a service provider's coverage area. In CSA architectures, customers receive their service as one of a number of customers within a geographical area that is served by a remote multiplexer, known as a *remote terminal* (RT). All customers on a remote terminal share access to the central office over a collection of shared wire pairs using traditional T-Carrier-like TDM.

The layout of a CSA is shown in Figure 1-27. The *central office terminal* (COT) is connected to the RT using a collection of standard 24-channel T-Carrier facilities. Because the T-Carrier is digital, it can be as long as required, using in-span repeaters to maintain signal quality. Thus, a CSA architecture can provide service to remote areas using significantly fewer wire pairs than would be required if each customer were given one. From the remote terminal, the local loops can run an additional 9,000 feet over 26-gauge wire or 12,000 feet over 24-gauge wire. Neither requires load coils over that distance.

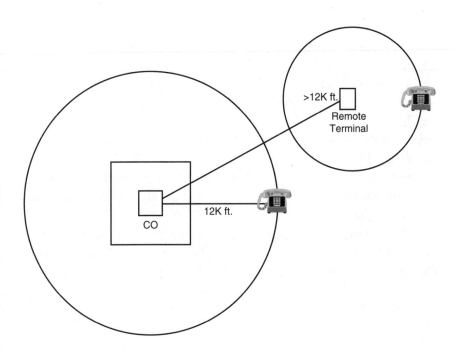

FIGURE 1-27 CSA architecture.

From a DSL point of view, traditional CSA presents a problem. T-Carrier allocates a single DS-0 channel to each customer, which effectively limits the maximum bandwidth achievable over that channel to 64 Kbps, far less than DSL requires. The good news is that the so-called *next generation digital loop carrier* (NGDLC) stands to overcome this challenge. NGDLC will rely largely on a fiber-based infrastructure, which overcomes the bulk of the available bandwidth problem; the channelization issue is remedied by the deployment of Bellcore Standard GR-303, which gives NGDLCs the capability to dynamically assign available bandwidth on an as-needed basis. Given GR-303's capability to provision bandwidth in increments greater than 64 Kbps and DSL's demand for high bandwidth, CSA architectures designed around NGDLC technology are well positioned to help bring about the widespread deployment of DSL.

LOOP LENGTH

DSL chipsets tend to be extremely sensitive to loop length. Most of the symmetric DSL services will operate at distances of 12,000 feet, while the asymmetric services will operate up to 18,000 feet across the loop. As long as deployment is limited to the local loop plant that is within the distance range stipulated by the DSL standards, the service stands a good chance of working. However, many of the existing chipsets are so distance sensitive that they will fail to work properly over loops that are as little as 650 feet longer than the stipulated maximum length.

Obviously, there is nothing that can be done to physically shorten a local loop. However, NGDLC technology may provide the solution, as will deployment of such technologies as SHDSL.

CROSSTALK AND BACKGROUND NOISE

Crosstalk can be minimized by ensuring to the greatest degree possible that only one type of DSL be deployed within a binder group[15] to minimize the amount of foreign-NEXT. Background noise is another matter entirely that, to a large extent, is unpredictable and therefore difficult to manage. Some electrical noise, such as that caused by electric motors, air conditioner and refrigerator compressors, and other RFI sources, can be managed through the judicious use of noise filters. Thermal noise on the other hand is caused by variations in ambient temperature, and short of refrigerating the cables, there is little that can be done to prevent it. It must, therefore, largely be worked around rather than eliminated.

DOWN THE ROAD

There is a significant amount of energy being funneled into DSL services and technology development. This will enable service providers to reach customers who are currently on the fringe of their service area for DSL. This is clearly a development to watch.

The integrated access devices market is also afire. In response to customer demand from smaller companies for the ability to move integrated voice and data over a single facility, a number of vendors have announced voice-over DSL solutions, while others have announced integrated access devices that deliver multiple lines and services across a single local loop. Using ATM as the transport mechanism, the so-called next-generation integrated access devices encapsulate IP-based data

---

[15]A binder group is a group of 25 cable pairs within a larger cable.

into ATM cells and provide the DSL interface for voice traffic. The *integrated access device* (IAD) manages all functions previously handled by a DSL modem as well as a bridge or router, and because it relies on ATM transport, it can take advantage of the granular QoS control that ATM enables. Once the traffic reaches the service provider's central office, a DSLAM routes data traffic to the data network and encapsulated voice traffic to a voice gateway where it is converted to a standard voice signal and handed off to a local switch.

When I wrote the first edition of this book in 1999 to 2000, *Data LECs* (DLECs) were poised to make a major contribution to the marketplace and become technological thorns in the backsides of the ILECs. For a while, they succeeded; then, in the beginnings of 2001, they began to falter. Three of them, Rhythms NetConnections, Covad Communications, and NorthPoint, succeeded with IPOs and as of August 1999 were valued in the $4 to $5 billion range on 1999 sales projections of $10 to $50 million. In January 1998, AT&T invested $25 million in Covad Communications, and MCI WorldCom invested $30 million in Rhythms NetConnections. Today, they are all but gone, having lost virtually 100 percent of their marketshare to the incumbent players.

DSL SUMMARY

From a service convergence perspective, DSL represents a major component in the evolving full-service network. Customer requirements are best met when the solution offered to them is a simple one that integrates all of their needs on a common infrastructure or platform. If service providers can overcome the technical issues that currently hinder the widespread deployment of DSL (and they certainly will), it will become a powerful force in the modern network and will ascend to the point of primacy in the battle for broadband access control. Furthermore, there is a management issue that must be addressed by would-be providers. It is far better to deny DSL service to a customer initially than to sell the service and later discover that it cannot be turned up on the customer's line because of inadequate loop performance. Service providers have therefore created processes to minimize the inconvenience to the customer by following a two-step approach. First, they test local loops that fall within the service area for potential DSL deployment using MLT and manual testing procedures to create databases that can be quickly consulted when a request for service comes in. Second, if the loop does qualify for DSL, they test the service carefully after turnup to ensure that the loop performs properly across the entire range of frequencies required by both the voice and data signal components. This process first ensures that the service is deliverable and, second, guarantees that it works when turned over to the customer, all without burdening the customer unnecessarily with technology barriers that they have no reason to be concerned with.

WIRELESS ACCESS TECHNOLOGIES

It is only in the last few years that wireless access technologies have advanced to the point that they are being taken seriously as contenders for the broadband local loop market. Traditionally, there was minimal infrastructure in place, and what there was bandwidth-bound and error-prone to the point that wireless solutions were not considered as serious contenders.

Wireless access technologies have undergone an evolution comprising three generations. *First-generation* (1G) systems, which originated in the late 1970s and throughout the 1980s, were entirely analog in nature and supported almost exclusively voice—very little data. They are characterized by the use of *frequency division multiple access* (FDMA) technology. In FDMA, users are assigned analog frequency pairs over which they send and receive. One frequency serves as a voice transmit channel, the other as a receive channel.

Second-generation (2G) systems, which came about beginning in the 1990s, were all-digital and were still primarily voice oriented, although data transport became more accepted. In second-generation systems, digital access became the norm through such technologies as *time division multiple access* (TDMA), *global system for mobile communications* (GSM), and *code division multiple access* (CDMA). In TDMA systems, the available frequency is broken into channels as in FDMA, but here the similarity ends. The channels are shared among a group of users who use a time division technique to ensure fair and equal access to the available channel pairs.

In CDMA, sometimes called *spread spectrum*, there is no channelization per se. Instead, all users share access to the entire range of available spectrum, and special techniques are utilized to isolate one conversation from another. These techniques include frequency hopping, in which each conversation hops randomly from one frequency to another and noise modulation, in which the signal is modulated against a pseudo-random noise signal so that it appears to be background noise to anyone listening in.

Finally, in 3G systems, we see the emergence of true broadband access and the promise of high-speed data in addition to voice. Access technologies for broadband will include *Wideband CDMA* (W-CDMA) or an enhanced version of GSM. Seventy-five percent of all digital cellular phones in the world are GSM-based (in fact, it is the most widely deployed access standard in the United States), and its popularity continues to grow. Cingular, VoiceStream, and AT&T Wireless offer GSM networks; Cingular has enhanced theirs with the *general packet radio service* (GPRS), which adds an always-on capability and high-speed packet access.

In late 1999, the ITU created a comprehensive set of 3G standards designed to accommodate the various technological directions that implementers have taken and to ensure that current systems can gracefully evolve to new 3G standards.

3G has undergone a series of significant evolutionary phases and deployment setbacks over the last couple of years, particularly in the United States, where a plethora of competing standards and a failure to deploy adequate spectrum has gotten in the way of technological progress. 3G actually comprises a collection of standards, including CDMA 2000, the *universal mobile telecommunications system* (UMTS), and W-CDMA. In a 2000 ruling, the ITU detailed requirements for 3G systems under a framework called *International Mobile Telephone* (IMT)-2000: data transmission capabilities at 144 Kbps inside a roaming vehicle and 2 Mbps to a fixed user, relying on packet transmission and offering global roaming.

As part of the original game plan, five terrestrial radio interface standards were accepted: IMT DS, also known as W-CDMA; IMT MC, also known as cdma2000; IMT TC, also known as UTRA TDD; IMT SC, also known as UWC-136 or EDGE; and finally, IMT FT, commonly known as DECT. Today, the key issue in 3G is spectrum availability. In the spring of 2000, the *World Radio Conference* (WRC) announced that the ranges of spectrum between 1710 and 1855 MHz, as well as 2520 to 2760 MHz, would be the worldwide allocations for 3G. Unfortunately, the U.S. Department of Defense already uses spectrum within that range and is reluctant to move. Some American carriers have simply decided to use their 2.5G spectrum allocations for the delivery of 3G services, adding technological tweaks to make it happen.

AT&T, Sprint, VoiceStream, and Cingular, the joint venture company formed by the Bellsouth and SBC wireless divisions, all own spectrum in the 1900-MHz range and are making plans to roll out 3G service or a 2.5G service that resembles the services available through broadband 3G. Sprint PCS and Verizon Wireless use CDMA and are making equally ambitious plans. The real winner today, though, is NTT DoCoMo, which rolled out the first 3G offering in October 2001. The service, called *Freedom of Mobile Multimedia Access* (FOMA), makes a wide range of services available to subscribers, including videoconferencing, Web access, and short messaging services, in addition to software download applications. The most popular downloads are custom ringing tones and screen logos, but other services are being formulated and will be offered soon. The hottest application currently is text messaging; the number of text messages sent globally each month grew from 4 billion in December 1999 to 20 billion in December 2000 and exceeded 40 billion in 2001. Revenue from the service currently exceeds $3 billion monthly and will exceed $5 billion by 2002.

FIXED WIRELESS TECHNOLOGIES

Recently, a new family of broadband wireless technologies has emerged that pose a serious threat to traditional wired access infrastructures. These include *local multipoint distribution service* (LMDS), MMDS, and *geosynchronous* (GEO) and *low earth orbit* (LEO) satellites.

LMDS

LMDS is a bottleneck resolution technology designed to alleviate the transmission restriction that occurs between high-speed LANs and WANs. Today, local networks routinely operate at speeds of 100 Mbps (Fast Ethernet) and even 1,000 Mbps (Gigabit Ethernet), which means that any local loop solution that operates slower than either of those poses a restrictive barrier to the overall performance of the system. LMDS offers a good alternative to wired options. Originally offered as CellularVision, it was seen by its inventor, Bernard Bossard, as a way to provide cellular television as an alternative to cable. In August of 1993, Bell Atlantic purchased an interest in the company and currently runs a multicell system in Brooklyn.

Operating in the 28-GHz range, LMDS offers data rates as high as 155 Mbps, the equivalent of SONET OC-3c. Because it is a wireless solution, it requires a minimal infrastructure and can be deployed quickly and cost effectively as an alternative to the wired infrastructure provided by incumbent service providers. After all, the highest cost component when building networks is not the distribution facility, but rather the labor required to trench it into the ground or build aerial facilities. Thus, any access alternative that minimizes the cost of labor will garner significant attention.

LMDS relies on a cellular-like deployment strategy under which the cells are approximately three miles in diameter. Unlike cellular service, however, users are stationary. Consequently, there is no need for LMDS cells to support roaming. Antenna/transceiver units are generally placed on rooftops as they need an unobstructed line of sight to operate properly. In fact, this is one of the disadvantages of LMDS (and a number of other wireless technologies): Besides suffering from unexpected physical obstructions, the service suffers from rain fade caused by absorption and scattering of the transmitted microwave signal by atmospheric moisture. Even some forms of foliage will cause interference for LMDS, so the transmission and reception equipment must be mounted high enough to avoid such obstacles—hence the tendency to mount the equipment on rooftops.

Because of its high-bandwidth capability, many LMDS implementations interface directly with an ATM backbone to take advantage of both its bandwidth and its diverse QoS capability. If ATM is indeed the transport fabric of choice, then the LMDS service becomes a broadband access alternative to a network capable of transporting a full range of services including voice, video, image, and data—the full suite of multimedia applications.

THE LMDS MARKET

In addition to Bell Atlantic, which bought an interest in CellularVision early on, other players have also stepped into the fray. Lucent, Broadband Networks, Texas Instruments, Bosch, Cisco, and Hewlett-Packard have all committed themselves to the game, and a number of service providers have adopted LMDS as their broadband wireless local loop solution. In July 1998, CellularVision agreed to sell 850 of its 1,300 LMDS licenses in New York City for $32.5 million to Winstar Communications, which planned to use them to deliver local telecommunications services within the greater New York City area. Although this seemed like a rather extreme step, consider the fact that as many as 80 percent of buildings in major cities are not served by fiber. This makes LMDS an ideal access scheme for broadband delivery.

Three key applications for LMDS have emerged in markets where the technology has been deployed: voice, remote LAN access and interconnection, and interactive television. These three applications alone make LMDS a powerful contender in the broadband access market.

Of course, fixed wireless has been hit hard, as have other segments of the telecomm space. The three major suppliers, Advanced Radio Telecom, Winstar, and Teligent, have all filed for bankruptcy, but continue to operate as they work their way through the restructuring process. All three took on enormous debts to build their networks quickly, but when the markets began to disappear in late 2000, they were left with no customers. Many customers were concerned about such issues as security and reliability, adding to the appeal of the companies' service offerings. Nevertheless, industry analysts predict that fixed wireless is real, will survive the turbulent market, and will see revenues exceed $14 billion by 2006.

MMDS

MMDS got its start as a wireless cable television solution. In 1963, a spectrum allocation known as the *Instructional Television Fixed Service* (ITFS) was carried out by the FCC as a way to distribute educational content to schools and universities. In the 1970s, the FCC established a two-channel metropolitan distribution service called the *Multipoint Distribution Service* (MDS). It was to be used for the delivery of pay-TV signals, but with the advent of inexpensive satellite access and the ubiquitous deployment of cable systems, the need for MDS went away.

In 1983, the FCC rearranged the MDS and ITFS spectrum allocation, creating 20 ITFS education channels and 13 MDS channels. In order to qualify to use the ITFS channels, schools had to use a minimum of 20 hours of airtime, which meant that ITFS channels tended to be heavily, albeit randomly, utilized. As a result, MMDS providers that use all 33 MDS and ITFS channels must be able to dynamically map requests for service to available channels in a completely transparent fashion, which means that the bandwidth management system must be reasonably sophisticated.

Because MMDS is not a true cable system (in spite of the fact that it has its roots in television distribution), there are no franchise issues for its use. (There are, of course, licensing requirements.) However, the technology is also limited in terms of what it can do. Unlike LMDS, MMDS is designed as a one-way broadcast technology and therefore does not typically enable upstream communication. Beginning in November 1997, Bellsouth deployed an MMDS-based home entertainment system in New Orleans, offering more than 160 channels of programming. MMDS enables transmission distances up to 35 miles, much farther than LMDS, and by December 1998, the company had a similar presence in Atlanta, Charleston, Birmingham, and Jacksonville as well. Today, their coverage area has expanded even more dramatically, and other companies have also entered the game. Sprint holds MMDS licenses in 90 markets and offers services in 14 areas.

However, many contend that there is adequate bandwidth in MMDS to provision two-way systems, which would make it suitable for voice, Internet access, and other data-oriented services. In fact, a number of providers have crafted two-way systems that are performing rather well. In May 1999, CAI Wireless, a wireless cable provider and the owner of MMDS-dependent CS Wireless Systems, announced their intent to be acquired by MCI Worldcom for approximately $475 million. CS Wireless became one of the largest users of MMDS in the world, operating in 11 markets. Unfortunately, wireless cable services provided by CAI Wireless and CS Wireless did not prove to be as profitable as had been expected, and the company filed for bankruptcy in 1998.

Clearly, companies like Sprint and MCI Worldcom see the value of wireless access to their own networks and have taken steps to bring about the technological convergence of innovative technologies to meet the growing customer demand for high-speed Internet access. The manufacturing sector certainly sees advantages in the LMDS and MMDS markets; more than a dozen companies now manufacture equipment for the two technologies, including such heavyweights as Lucent Technologies, Nortel Networks, and Newbridge Networks, to name a few. Furthermore, both Cisco and Motorola have thrown their hats in the ring; in June 1999, the two companies announced their joint intent to acquire Bosch Telecom's LMDS division.

SO WHAT'S THE MARKET?

Service providers looking to sell LMDS and MMDS technologies into the marketplace should target small and medium-size businesses that experience measurable peak data rates and that are looking to move into the packet-based transport arena. Typical applications for the technologies include LAN interconnections, Internet access, and cellular backhauls between *mobile telephone switching offices* (MTSOs) and newly deployed cell sites. LMDS tends to offer higher data rates than MMDS; LMDS peaks out at a whopping 1.5 Gbps, while MMDS can achieve a maximum transmission speed of about 3 Mbps. Nevertheless, both have their place in the technology pantheon.

SATELLITE TRANSMISSION

It can be observed that one orbit, with a radius of 42,000 km, has a period of exactly 24 hours. A body in such an orbit, if its plane coincided with that of the earth's equator, would revolve with the earth and would thus be stationary above the same spot on the planet. It would remain fixed in the sky of a whole hemisphere, and unlike all other heavenly bodies, would neither rise nor set. A body in a smaller orbit would revolve more quickly than the earth and so would rise in the West, as indeed happens with the inner moon of Mars . . .

Let us now suppose that such a station were built in this orbit that could be provided with receiving and transmitting equipment (the problem of power will be discussed later) and could act as a repeater to relay transmissions between any two points on the hemisphere beneath, using any frequency that will penetrate the ionosphere. If directive arrays were used, the power requirements would be very small, as a direct line of sight transmission would be used. There is the further important point that arrays on the earth, once set up, could remain fixed indefinitely. Moreover, a transmission received from any point on the hemisphere could be broadcast to the whole of the visible face of the globe, and thus the requirements of all possible services would be met (see Figure 1-28).

In October 1945, Arthur C. Clarke published a paper in *Wireless World* entitled, "Extra-Terrestrial-Relays: Can Rocket Stations Give World-Wide Radio Coverage?" In his paper, Clarke proposed the concept of an orbiting platform that would serve as a relay facility for radio signals sent to it that could be turned around and retransmitted back to Earth with far greater coverage than was achievable through terrestrial transmission techniques. His platform would orbit at an altitude of 42,000 kilometers (25,200 miles) above the equator where it would orbit at a speed identical to the rotation speed of the earth. As a consequence, the satellite would appear to be stationary to Earth-bound users.

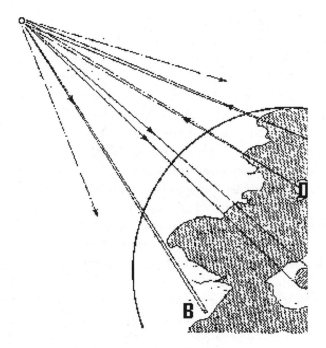

FIGURE 1-28   Excerpted from *Wireless World*, October 1945.

Satellite technology may prove to be the primary communications gateway for regions of the world that do not yet have a terrestrial wired infrastructure, particularly given the fact that they are now capable of delivering broadband services. In addition to the United States, the largest markets for satellite coverage are Latin America and Asia, particularly Brazil and China. According to the Strategis Group's second edition of *World Mobile Satellite Telephony Markets: 1999–2007*, the Asia-Pacific region will develop the largest base of subscribers with more than 30 percent of the global total. Most of them will be mobile users, although the fixed wireless market will grow dramatically, with Latin America having 35 percent of the rural fixed wireless user base.

GEO SATELLITES

Clarke's concept of a stationary platform in space forms the basis for today's geostationary or GEO satellites. Ringing the equator like a string of pearls, these birds provide a variety of services, including 64-Kbps voice, broadcast television, video-on-demand services, broadcast and interactive data, and point-of-sale applications, to name a few. Although satellites are viewed as technological marvels, the real magic lies more with what it takes to harden them for the environment in which they must operate and what it takes to get them there than it does their actual operational responsibilities. Satellites are, in effect, nothing more than a sophisticated collection of assignable, on-demand repeaters—in a sense, the world's longest local loop.

From a broadcast perspective, satellite technology has a number of advantages. First, its one-to-many capabilities are unequaled. Information from a central point can be transmitted to a satellite in geostationary orbit; the satellite can then rebroadcast the signal back to Earth, covering an enormous service footprint. Consider, for example, TCI's *Headend in the Sky* (HITS) model. From the company's National Digital Television Center in Littleton, Colorado, TCI (now part of AT&T Broadband) can generate a remarkable variety of programming content, which is digitized, compressed, and uplinked to a constellation of satellites. The satellites then rebroadcast the content to every headend in the system, making it possible for the tiniest cable system in the company to offer the same channel lineup as the largest. Similarly, direct broadcast satellite companies like DirecTV can offer equally robust channel lineups to their far-flung customers.

Because the satellites appear to be stationary, the earth stations actually *can* be. One of the most common implementations of GEO technology is seen in the *very small aperture terminal* (VSAT) dishes that have sprung up like mushrooms on a summer lawn. These dishes are used to provide both broadcast and interactive applications; the small DBS dishes used to receive TV signals are examples of broadcast applications, while the dishes seen on the roofs of large retail establishments, automobile dealerships, and convenience stores are typically (although not always) used for interactive applications, such as credit card verification, inventory queries, e-mail, and other corporate communications. Some of these applications use a satellite downlink, but rely on a telco return for the upstream traffic.

One disadvantage of GEO satellites has to do with their orbital altitude. On the one hand, because they are so high, their service footprint is extremely large. On the other hand, because of the distance from the Earth to the bird, the typical transit time for the signal to propagate from the ground to the satellite (or back) is about half a second, which is a significant propagation delay for many services. Should an error occur in the transmission stream during transmission, the need to detect the error, ask for a retransmission, and wait for the second copy to arrive could be catastrophic for delay-sensitive services like voice and video. Consequently, many of these systems rely on forward error correction transmission techniques that enable the receiver to not only detect the error, but correct it as well.

The biggest supplier in the country for VSAT and other GEO services is Hughes, although there are others, including GTE SpaceNet, Inmarsat, Intelsat, and RCA.

LOW/MEDIUM EARTH ORBIT SATELLITES (LEO/MEO)

In addition to the GEO satellite arrays, there are a variety of lower orbit constellations deployed known as *low and medium earth orbit satellites* (LEO/MEO). Unlike the GEO satellites, these orbit at lower altitudes—400 to 600 miles, far lower than the 23,000-mile altitude of the typical GEO bird. As a result of their lower altitude, the transit delay between an Earth station and a LEO satellite is virtually nonexistent. However, another problem exists with LEO technology. Because the satellites orbit pole to pole, they do not appear to be stationary, which means that if they are to provide uninterrupted service, they must be able to hand off a transmission from one satellite to another before the first bird disappears below the horizon. This has resulted in the development of sophisticated satellite-based technology that emulates the functionality of a cellular telephone network. The difference is that in this case, the user does not appear to move; the cell does!

IRIDIUM   Perhaps the best known example of LEO technology is Motorola's ill-fated Iridium deployment. Comprising 66 satellites[16] in a polar array, Iridium was designed to provide voice service to any user, anywhere on the face of the Earth. The satellites would provide global coverage and would hand off calls from one to another as the need arose. Unfortunately, Iridium's marketing strategy was flawed; their prices were high, their phones large and cumbersome (one newspaper article referred to them as manly phones), and their market significantly overestimated. Additionally, their system was only capable of supporting 64-Kbps voice services, a puny bandwidth allocation in these days of customers with broadband desires.

---

[16]The system was named Iridium because in the original design, the system was to require 77 satellites, and 77 is the atomic number of that element. Shortly after naming it Iridium, however, the technologists in the company determined that they would only need 66 birds. They did not rename the system Dysprosium.

Iridium is not alone. ICO Global Communications, another satellite-based global communications company, filed for bankruptcy in August 1999. In November of that same year, Craig McCaw of Teledesic and Eagle River offered salvation for the company with an investment package valued at as much as $1.2 billion.

GLOBALSTAR   Others have been slightly more successful. Globalstar's 48-satellite array offers voice, short messaging, roaming, global positioning, fax, and data transports up to 9600 bps, and although the data rates are miniscule by comparison to other services, the converged collection of services they provide is attractive to customers who want to reduce the number of devices they must carry with them in order to stay connected. Unfortunately, Globalstar has also suffered major financial losses in the marketplace, and many analysts now question their viability along with that of its competitors.

ORBCOMM   This is a partnership jointly owned by Orbital Sciences Corporation and Teleglobe Canada. Their satellites are nothing more than extraterrestrial routers that interconnect vehicles and earth stations to facilitate the deployment of such packet-based applications as two-way short messaging, e-mail, and vehicle tracking. Their constellation has a total of 35 satellites.

TELEDESIC   After Iridium, the best-known LEO satellite services company is Teledesic. Started in 1990 by Craig McCaw and Bill Gates, Teledesic will comprise an array of 288 satellites capable of providing up to 64 Mbps downstream and 2 Mbps upstream for voice, videoconferencing, and data, with plans in place to offer symmetric 64-Mbps service sometime in the future. In 1998, Motorola joined the Teledesic team, and in 1999, the company signed a launch agreement with Lockheed Martin. They plan to be fully operational by 2004.

In Summary

As a service provisioning technology, satellites may seem so far out (no pun intended) that they may not appear to pose a threat to more traditional telecommunications solutions. At one time, that is, before the advent of LEO technology, this was largely true. GEO satellites were extremely expensive, offered low bit rates, and suffered from serious latency that was unacceptable for many applications.

Today, this is no longer true. Some GEO satellites offer high-quality, two-way transmission for certain applications. LEO technology has advanced to the point that it now offers low-latency, two-way communications at broadband speeds, is relatively inexpensive, and, as a consequence, poses a clear threat to terrestrial services. On the other hand, the best way to eliminate an enemy is to make the enemy a friend. Many traditional service providers have entered into alliances with satellite providers; consider the agreements that exist between DirecTV, a high-quality, wireless alternative to cable, and Bell Atlantic, GTE, Cincinnati Bell, and SBC corporations. Again, there is convergence at work here. Customers have indicated repeatedly that they like the idea of being able to go to a single source for all of their communications needs. By joining forces with satellite providers, the ILECs create something of a market block that will help them stave off the incursion of cable. Between the minimal infrastructure required to receive satellite signals and the soon-to-be ubiquitous deployment of DSL over twisted pair, incumbent local telephone companies and their alliance partners are in a reasonably good position to counter the efforts of cable providers wanting to enter the local services marketplace.

Other players continue to make plans to enter the market as well. Astrolink International, funded largely by Lockheed Martin, plans to launch a GEO satellite in early 2003. Tachyon offers 256-Kbps upload with T-1 download speeds, while StarBand Communications offers 500-Kbps downstream with 128-Kbps upstream. Similarly, Alcatel's SkyBridge, potentially Teledesic's biggest competitor, plans to begin service in 2004.

THE SPECIAL CASE OF
PREMISES ACCESS

It would be improper to discuss access technologies without also discussing the technologies used to *access* the access technologies—LANs and variations on the LAN theme.

LANs have traditionally fallen into two primary categories characterized by the manner in which they access the shared transmission medium (shared among all the devices on the LAN). The first and most common is called *contention,* and the second group is called *distributed polling.*

CONTENTION-BASED LANS

In contention-based LANs, devices attached to the network vie for access using the technological equivalent of gladiatorial combat. If it feels good, do it is a good way to describe the manner in which they share access. If a station wants to transmit, it simply does so, knowing that there exists the possibility that the transmitted signal may collide with that generated by another station that transmits at the same time. In the event that a collision occurs, both stations back off, wait a random amount of time, and try again. Ultimately, each station *will* get its turn, although how long they have to wait is based upon how busy the LAN is. These systems are characterized by what is known as *unbounded delay* because there is no upward limit on how much delay a station can incur as it waits to use the shared medium.

The protocol that these LANs usually employ is called *Carrier Sense Multiple Access with Collision Detection* (CSMA/CD). In CSMA/CD, stations observe the following guidelines when attempting to use the shared network. First, they listen to the shared medium to determine whether it is in use or not—that's the carrier sense part. If the LAN is available, they begin to transmit, but continue to listen while they are transmitting, knowing that another station could also choose to transmit at the same time—that's the multiple access part. In the event that a collision is detected, usually indicated by a dramatic increase in the signal power measured on the shared LAN, both stations back off and try again. That's the collision detection part.

Ethernet is the most common example of a CSMA/CD LAN. Originally released as a 10-Mbps product based on IEEE standard 802.3, Ethernet rapidly became the most widely deployed LAN technology in the world. As bandwidth-hungry applications such as e-commerce, *enterprise resource planning* (ERP), and Web access evolved, transport technologies advanced, and bandwidth availability (and capability) grew, 10-Mbps Ethernet began to show its age.

The other aspect of the LAN environment that began to show weaknesses was the overall topology of the network itself. LANs are broadcast environments, which means that when a station transmits, every station on the LAN segment hears the message (see Figure 1-29). Although this is a simple implementation scheme, it is also wasteful of bandwidth because stations hear broadcasts that they have no reason to hear. In response to this, a technological evolution occurred. It was obvious to LAN implementers that the traffic on most LANs was somewhat domain-oriented, that is, it tended to cluster into communities of interest based on the work groups using the LAN. For example, if employees in sales shared a LAN with shipping and order processing, three discernible traffic group-

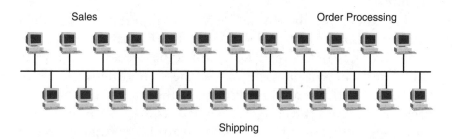

FIGURE 1-29   Three departments, single LAN.

ings emerged according to what network architects call the 80:20 Rule. The 80:20 Rule simply states that 80 percent of the traffic that originates in a particular work group tends to stay in that work group, an observation that makes network design distinctly simpler. If the traffic naturally tends to segregate itself into groupings, then the topology of the network could change to reflect those groupings. Thus was born the bridge.

Bridges are devices with two responsibilities: They filter traffic that does not have to propagate in the forward direction and they forward traffic that does. For example, if the network described previously were to have a bridge inserted in it (see Figure 1-30), all of the employees in each of the three work groups would share a LAN segment, and each segment would be attached to a port on the bridge. When an employee in sales transmits a message to another employee in sales, the bridge is intelligent enough to know that the traffic does not have to be forwarded to the other ports. Similarly, if the sales employee now sends a message to someone in shipping, the bridge recognizes that the sender and receiver are on different segments and thus forwards the message to the appropriate port, using address information in a table that it maintains (the filter/forward database).

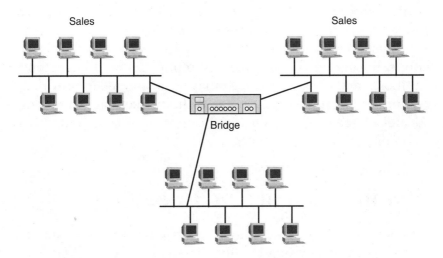

FIGURE 1-30   Three departments, LAN segmented with bridge.

Following close on the heels of bridging is *LAN switching*. LAN switching qualifies as bridging with an attitude. In LAN switching, the filter/forward database is distributed—that is, a copy of it exists at each port, which implies that simultaneous traffic-handling decisions can be made by different ports. This enables the LAN switch to implement full-duplex transmissions, reduce overall throughout delay, and, in some cases, implement per-port rate adjustments. The first 10-Mbps Ethernet LAN switches emerged in 1993, followed closely by Fast Ethernet (100-Mbps) versions in 1995 and Gigabit Ethernet (1000-Mbps) switches in 1997. Fast Ethernet immediately stepped up to the marketplace bandwidth challenge and was quickly accepted as the next generation of Ethernet, with per-port prices dropping 47 percent between 1997 and 1998 in concert with a four-fold increase in port sales during the same timeframe. To date, the prices have continued to decline, and the solution has caught on with the SoHo market as well as the enterprise environment.

Gigabit Ethernet is still in a somewhat nascent stage, but most analysts believe that it will experience a similar uptake rate. Dataquest predicts that Gigabit Ethernet sales will grow to $2.5 billion by year-end 2002; this is a reasonable number, considering that 2 million ports were sold in 1999 with expectations of hitting a total installed base of 18 million by EOY 2002. Emerging applications certainly make the case for Gigabit Ethernet's bandwidth capability: LAN telephony, server interconnections, and video to the desktop all demand low-latency solutions, and Gigabit Ethernet may be positioned to provide it. A number of vendors have entered the marketplace, including Alcatel, Lucent Technologies, Nortel Networks, and Cisco Systems.

DISTRIBUTED POLLING LANS

In addition to the gladiatorial combat approach to sharing access to a transmission facility, there is a more civilized technique known as *distributed polling* or, as it is more commonly known, *token passing*. IBM's Token-Passing Ring is perhaps the best known of these products, followed closely by *fiber distributed data interface* (FDDI), a 100-Mbps version often seen in campus and metropolitan area networks.

In token-passing LANs, stations take turns with the shared medium, passing the right to use it from station to station by handing off a token that gives the bearer the one-time right to transmit while all other stations remain quiescent. This is a much fairer way to share access to the transmission medium than CSMA/CD because although every station has to wait for its turn, it is absolutely guaranteed that it will get that turn. These systems are, therefore, characterized by bounded delay because there is a maximum amount of time that any station will ever have to wait for the token.

Traditional token ring LANs operate at two speeds—4 and 16 Mbps. Like Ethernet, these speeds were fine for the limited requirements of text-based LAN traffic that was characteristic of early LAN deployments. However, as demand for bandwidth climbed, the need to eliminate the bottleneck in the token ring domain emerged, and fast token ring was born. In 1998, the IEEE 802.5 committee (the oversight committee for token ring technology) announced draft standards for 100-Mbps high-speed token ring (HSTR 802.5t). A number of vendors stepped up to the challenge and began to produce high-speed token ring equipment, including Madge Networks, IBM, and Olicom (which has since sold its token ring business to Madge).

Gigabit token ring is on the horizon as draft standard 802.5v, and although it has not yet emerged as a widespread access technology, it may as the demand for bandwidth grows. However, token rings tend to be more expensive than Ethernet, and given the widespread deployment and overwhelming global success of fast and gigabit Ethernet, high-speed token rings may have a battle before it. To date, there are no known commercial deployments, which means that the solution may be relegated to the status of an academic (albeit valuable) exercise.

WIRELESS ETHERNET: 802.11

In the last year, wireless LANs have grown in importance because of the increasing demand for mobility in both the enterprise and SoHo environments. One solution that has emerged is 802.11, a family of specifications for *wireless LANs* (WLANs) developed by an IEEE working group. The *working group* (WG) has defined four specifications: 802.11, 802.11a, 802.11b, and 802.11g. All four use the Ethernet protocol and *Carrier Sense Multiple Access with Collision Avoidance* (CSMA/CA) for sharing access to the available wireless bandwidth.

The 802.11b standard, sometimes called *Wi-Fi*, was designed to be fully compatible with traditional 802.11, both technologically and philosophically. The modulation used in 802.11 has historically been *phase-shift keying* (PSK), while the modulation method selected for 802.11b is called *complementary code keying* (CCK). The technique enables the delivery of higher data speeds and is less susceptible to interference from multipath propagation.

The 802.11a specification is used primarily in wireless ATM systems and is deployed in architectures that rely on access hubs. 802.11a operates at frequencies between 5 and 6 GHz. It uses a modulation scheme called *orthogonal frequency-division multiplexing* (OFDM), which facilitates data transports at speeds as high as 54 Mbps. Today, communication is available at 6, 12, or 24 Mbps, with 54 promised in the future.

The most recent standard, 802.11g, provides service over relatively short distances at up to 54 Mbps compared to the 11 Mbps of 802.11b. Like 802.11b, 802.11g operates in the 2.4-GHz range and is therefore fully compatible with it.

A related technology is HiperLAN. HiperLAN is a *wireless LAN* (WLAN) standard that is primarily used in Europe. There are two key specifications, HiperLAN/1 and HiperLAN/2. Both have been adopted by the ETSI and provide features and capabilities similar to those of 802.11 used in the United States and other countries. HiperLAN/1 provides transports of up to 20 Mbps in the 5-GHz range of the spectrum, while HiperLAN/2 operates at up to 54 Mbps. HiperLAN/2 is compatible with 3G systems for the transmission of data, images, and voice communications. It is intended for worldwide implementation.

THE HOME PHONELINE NETWORKING ALLIANCE (HOMEPNA)

HomePNA is a not-for-profit association founded in 1998 by 11 companies to bring about the creation of uniform standards for the deployment of interoperable in-home networking solutions. The companies, 3Com, AMD, AT&T Wireless, Compaq, Conexant, Epigram, Hewlett-Packard, IBM, Intel, Lucent Technologies, and Tut Systems, are committed to the design of in-home communications systems that will enable the transport of LAN protocols across standard telephone inside wire. On April 9, 2001, HomePNA announced approval by the ITU of a technical specification for a worldwide home phoneline networking standard. The Recommendation G.989.1 (Phoneline Networking Transceivers—Foundation) covers specifications for home phoneline networking and details the key characteristics of devices designed for the transmission of data over existing phoneline wiring within the home. HomePNA-certified products, including preconfigured PCs, NICs, broadband modems, gateways, and home networking chip solutions, are designed to adhere to the new ITU G.989.1 Recommendation.

Currently, HomePNA operates at 1 Mbps and enables the user to employ every RJ-11 jack in the house as both telephone and data ports for the interconnection of PCs, peripherals, and telephones. The standard enables interoperability with existing access technologies, such as xDSL and ISDN, and therefore provides a good migration strategy for SoHo and telecommuter applications. The initial deployment utilizes Tut Systems' HomeRun product, which enables telephones and computers to share the same wiring and to simultaneously transmit. Extensive tests have been performed by the HomePNA, and although some minor noise problems caused by AC power noise and telephony-coupled impedance have resulted, they have been able to eliminate the effect through the use of low-pass filters between the devices causing the noise and the jack into which they are plugged.

BLUETOOTH

Much has been said in recent months about the on-again, off-again futures of Bluetooth. The Bluetooth standard, named for tenth-century Danish king Harald Bluetooth who united Denmark and part of Norway into a single kingdom, is a standard developed by a group of manufacturers that enables any device, including computers, cell phones, and headphones, to automatically establish wireless device-to-device connections. The Bluetooth Special Interest Group, comprising more than 1,000 companies, works to support Bluetooth's role as a replacement technology for connectivity between peripherals, telephones, and computers.

Bluetooth works at two levels. First, it provides wireless protocol agreement at the physical level. It also provides agreement at the data link level, where products have to agree on transmission schemes and communication control rules to be sure that the transmitted message arrives properly.

Bluetooth operates at 2.45 GHz, a spectrum allocation that has been set aside for the use of *industrial, scientific and medical devices* (ISM). The system is capable of transmitting data at 1 Mbps, although headers and handshaking information consume about 20 percent of this capacity. A number of other devices operate within this domain as well, including baby monitors, garage-door openers, and some cordless phones. The process of ensuring that Bluetooth and these other devices don't interfere with one another has been a crucial part of the design process.

Applications suggested for Bluetooth are quite interesting, and although they are distance limited because of the technology's low transmit power level, plenty of opportunities have presented themselves. The standard specifies operational distances of up to 100 meters between Bluetooth-equipped devices, but most will only be able to operate within 10 meter radii because of low transmit power. Bluetooth is designed for the creation of *personal area networks* (PANs) that rely on picocells for connectivity. As a result, all sorts of applications are possible. Bluetooth-equipped PDAs could pay toll charges on the freeway as the car passes through or pay parking meters, gas stations, or the carwash. Bluetooth could provide connectivity between a PC and its various peripherals or between a cellphone and a PDA. Any application that requires short-distance, high-bandwidth connectivity could be implemented over a Bluetooth link.

WAP

An alternative solution is WAP. Originally developed by Phone.com, WAP has proven to be a disappointment for the most part. Because it is designed to work with 3G wireless systems and because 3G systems have not yet materialized, some have taken to defining WAP to mean "wrong approach to portability." Germany's D2 network reports that the average WAP customer uses it less than two minutes per day, which is tough to make money on when service is billed on a usage basis. 3G will be the deciding factor; when it succeeds, WAP will succeed —unless 802.11's success continues to expand.

THE MOBILE APPLIANCE

The mobile appliance concept is enjoying a significant amount of attention of late because it promises to herald in a whole new way of using network and computer resources—*if it works as promised.* The problem with so many of these new technologies is that they overpromise and underdeliver—precisely the opposite of what they're supposed to do for a successful rollout. 3G, for example, has been billed as the wireless Internet. Largely as a result of that billing, it has failed. It is *not* the Internet—far from it. The bandwidth isn't there, nor is a device that can even begin to offer the kind of image quality that Internet users have become accustomed to. Furthermore, the number of screens that a user must go through to reach a desired site (I have heard estimates as high as 22!) is far too high. Therefore, until the user interface, content, and bandwidth challenges are met and satisfied, the technology will remain exactly that—a technology. There is no application yet, and *that's* what people are willing to pay money for.

ACCESS TECHNOLOGIES SUMMARY

This is a book about convergence, and as such we must consider how the convergence phenomenon affects the access technologies discussed in this section. We have examined a variety of options for connecting the customer to the wide area transport network: ISDN, T-1, cable modems, 56K modems, xDSL, wireless solutions (such as LMDS, MMDS, and satellite), and premises schemes (such as LANs and HomePNA technology). In all of these we have seen one recurring theme: the evolution of technology in each case to satisfy the demand for network bandwidth that will meet the needs of ever-evolving applications required by the customer to achieve a competitive advantage in their marketplace.

Remember the convergence mantra: Discrete knowledge about the customer's business goals and activities translates into a choice of technologies. The customer will ask for the best access solution possible; the access provider must respond with the appropriate collection of technologies and must not burden the customer with how those underlying technologies work. Ideally, what the customer sees is not the bits and bytes of technology, but rather a solution to a business problem.

Each of the access technologies described fits a particular niche; the key to success lies in correctly matching a unique technology to a specific solution. ISDN, for example, is the target of criticism and even ridicule in some circles, but it *does* offer 128-Kbps service *right now* in the areas where it is available, something 56-Kbps modems can't begin to achieve. Although DSL seems like the ideal technology for most problems, it is far from universally available and still suffers from vexing problems that stymie its widespread deployment in the near-term. DSL will certainly overcome its problems, but in the meantime, cable modems stand ready to fill the broadband access void with their own flavor of high-speed access—as well as content. Unfortunately, cable modems face the same spotty deployment challenges that plague the DSL providers, not to mention that cable is nowhere near as widely deployed as twisted pair. There is no question that the ILECs have a significant advantage with their switched local loop plant, but companies like AT&T are closing the gap by forming alliances with cable providers like TCI. They understand that the magic formula in the game has more than one ingredient. Remember the words of Claudio Monasterio of Lucent Technologies, "The technology is nothing more than a detail."

Service providers must take heed of their title and role in the converging marketplace. They are and are expected to be *service providers*. The difficult part of that expectation is the required evolution from being an access and transport provider to being a provider of customer solutions based on the technologies they understand, manage, and sell. As we will see in a later section, service providers must learn to focus not exclusively on their immediate customer but on the needs of their *customer's customer*. After all, the immediate customer will use whatever solution is provided to them to create solutions to their own customers' business challenges. If service providers take the initiative to look beyond their immediate market to that third tier, make an effort to understand the issues faced there, and craft solutions that will satisfy both the second- and third-tier business concerns, they win the game. This is not a game of who has the best technology; it is a game of who provides the best service. If technology happens to be the basis for the service, so be it. The customer, however, shouldn't have to care.

TRANSPORT TECHNOLOGIES

Transport technologies, usually represented as the famous cloud seen more or less ubiquitously in telecommunications product and service literature, comprise both dedicated private-line facilities and switched technologies, all designed to move traffic from one access point to another at the highest speed possible.

Traditionally, private-line facilities were the most common technique, with bandwidth ranging from as little as 300 bps in the early days of data transmission to the almost unimaginable Gb speeds of the North American SONET and its global counterpart, the *Synchronous Digital Hierarchy* (SDH). Add to those the remarkable bandwidth multiplication capabilities provided by *wavelength division multiplexing* (WDM) and we soon find that bandwidth is no longer the bottleneck that it once was. In fact, there is so much bandwidth available today that for the first time, commodity markets are appearing where it can be purchased like any other commodity—soy beans, pork bellies, Louisiana sweet crude, and bandwidth.

This poses a threat to those companies that have traditionally made their profits through the sale of bandwidth as a limited resource. As companies like Qwest and Level 3 build out their optical networks and bill themselves as the carriers' carriers, a once-scarce resource now becomes so widely available that the price for it plummets to near zero. When a product becomes so widely available from multiple suppliers that the only way to differentiate between them is on price alone, the product officially becomes a commodity. Obviously, service providers do not want to play in the commodity game, but if they view themselves as nothing more than access and transport providers, that's what they will become. Keep in mind, however, that the *good* news about being a commodity provider is that by definition, everyone wants your product. To maintain competitive advantage, service providers must heed the words of Harvard business professor Michael Porter, author of numerous books on competitive advantage and strategy: "A firm differentiates itself from its competitors when it provides something unique that is valuable to buyers beyond simply offering a low price."[17] In the access and transport game, everyone has a network, and by and large, they are all equally capable. Therefore, having the best network or the most expensive technology does not give service providers an advantage because the network itself is a commodity. Traditional access and transport companies must, therefore, become more than merely a data transport company. They must overlay a set of services on their physical networks that changes them from commodity providers into *service* providers. At that moment, the underlying technology becomes a valuable piece of the overall package.

During the early 1980s, when voice was the predominant traffic component on service provider networks, it was occasionally stated that if we build a voice network, the data can ride for free. This observation made perfect sense. Data transmission at that time represented a tiny component of the overall traffic volume. Not only that, it soon became a fundamental component of the competitive voice services market. With the implementation of the SS7 data network and its dedicated databases, it soon became possible to offer so-called intelligent network applications, which became enormous moneymakers for service providers. Consider the following example: A customer with a flat-rate monthly service receives a telephone call, but doesn't get to the phone before the caller hangs up. The customer examines their Caller ID display (for which they pay an additional monthly fee) and decides they want to speak with the caller. They dial *69, which invokes an SS7 service that automatically calls the caller back—for an additional fee. While they are on the phone, the customer hears a tone in his or her ear, indicating another incoming call that they then answer while putting the other person on hold—all for a fee. At some point during the call, they add on a third party using three-way calling, again, for a fee.

---

[17]Porter, Michael. *Competitive Advantage: Creating and Sustaining Superior Performance.* New York: The Free Press, 1985, page 120.

Please note that in this scenario, there is no charge associated with the duration of the call itself other than the low, flat-rate monthly fee. However, this simple phone call, because of the added data services that SS7 provides, generates approximately $6 in added revenue. By adding data services on top of the voice network's simple transport capabilities, *data become the killer application for voice.*

Fast forward: Voice is no longer the predominant traffic component on modern networks, yet it still accounts for the majority of the revenue. In August 1998, SBC Executive Vice President for Corporate Planning Michael Turner observed that 98 percent of the company's revenue derived from circuit-switched voice communications and went on to observe that even when data become more than half of the traffic SBC transports across their network, it will still account for less than half of their revenue. Even today, voice is the major source of revenue, as expressed in this quote from Jim Crowe, founder and CEO of Level 3 Communications:[18]

"Data is, by bit count, some 50 percent of the traffic flow through networks. Yet voice still represents 91 cents of each dollar billed to customers because telecommunications companies charge 15 times more for data than they do for voice. So there's a huge opportunity to profit based on the telecom market's shift to carrying data. But voice's revenue-generating advantage in billing will not disappear overnight. So beware of providers who concentrate on the multimedia data market and ignore the voice market. They could go broke if the multimedia market fails to grow quickly enough."

So although data have turned out to be the killer application for voice networks, service providers are now faced with the challenge of determining what the killer application for data networks will be as the world migrates to a converged access and transport infrastructure. Many believe, ironically, that the killer application for data networks will be—voice.

---

[18]"Telecom Shakeup Offers IT Opportunities." *IEEE ITPro*, May/June 1999.

CLOUDS

So why do we draw clouds to represent the transport network? The typical answers are

- We don't know what's in there.
- We don't *want* to know what's in there.
- There's no *need* to know what's in there.
- It doesn't *matter* what's in there.
- Clouds are easier to draw than hairballs.

To a large extent, all of the aforementioned reasons are true. In many cases we *don't* know what's in the cloud because we've never had to know. In the days prior to divestiture, when there was only one game in town from which to choose (called AT&T), the service selection scenario might have gone something like this:

> *Customer*: "Look, I've got a big problem. I'm a large bank with branches all over town. I have a mainframe computer in the data center downtown. I've got tellers in all the branches with dumb terminals that have to talk with it; I've got a few ATM machine scattered here and there, and —-" (interrupted)
>
> *AT&T Salesperson*: "Shhhh . . . Look, don't worry about the stuff in the middle. You just tell us where each location is and what they do there, and we'll take care of the stuff in the middle. *You take care of your banking operations, and let us take care of the network —that's our job."*

In this scenario, the network sales person did not really know or *want* to know what was in the cloud, nor did the customer. The important thing is that there was no *need* for either one to know because there were plenty of highly technical people in the phone company's ranks who would see to it that the bank got what they needed for continuing business operations. The AT&T person was free to concentrate on the customer's business issues.

That takes care of the first three bullets in the list. The fourth, *it doesn't matter what's in the cloud,* is also true. In the ideal world of service provisioning, the customer should have no reason to care what's in there. All they should have to care about is that they are given a gozinta and a gozouta (see Figure 1-31) that disappear into the fluffy opaqueness of the cloud, and as long as they connect to them correctly, their voice and data needs will be met, end of story.

This scenario represents the ideal world of service provisioning. If providers build their networks with the customers' requirements in mind (knowledge) and therefore choose the right technologies, they can sell access to a multipurpose cloud that satisfies *all* of the customer's network requirements. They become the provider of choice because there is no reason for the customer to go anywhere else. The service provider delivers exactly what the customer requires, where and when they need it. Furthermore, they may form strategic alliances with providers of other desirable services, such as content, as demon-

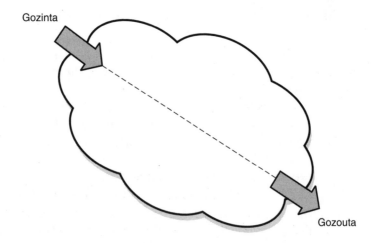

FIGURE 1-31   The network cloud.

strated by the relationships that exist between AOL, Prodigy, and the ILECs. They are no longer merely access and transport companies; they are true *service* providers.

THE EVOLUTION OF TRANSPORT

In this section we will examine the principal transport technologies, beginning with dedicated facilities such as private line, SONET/SDH, and microwave. Then we will continue with the evolution of switched services from traditional circuit and packet switching to modern fast packet architectures. The section will conclude with routing and the special case of the IP, the Rosetta Stone of communications technologies.

POINT-TO-POINT ARCHITECTURES

Point-to-point technologies do exactly what their name implies: They connect one point directly with another. For example, it is common for two buildings in a downtown area to be connected by a point-to-point microwave or infrared circuit because the cost of establishing it is far lower than the cost to put in physical facilities in a crowded city. Many businesses rely on dedicated, point-to-point optical facilities to interconnect locations, especially businesses that require dedicated bandwidth for high-speed applications. Of course, point-to-point does not necessarily imply high bandwidth; many locations use 1.544-Mbps T-1 facilities for interconnections, and some rely on lower-speed circuits where higher bandwidth is not required. In the last 18 months, metropolitan area networking has become a significant focal point for service providers as corporate models have changed and with those changes a need for low-cost, high-bandwidth transport between corporate facilities within a metro area has risen.

Dedicated facilities provide bandwidth from as low as 2400 bps to as high as multiple Gbps. 2400-bps analog facilities are not commonly seen, but are often used for alarm circuits and telemetry, while circuits operating at 4800 and 9600 bps are used to access interactive host-based data applications.

Higher-speed facilities are usually digital and are often channelized by dedicated multiplexers and shared among a collection of users or by a variety of applications. For example, a high-bandwidth facility that interconnects two corporate locations might be dynamically subdivided into various-sized channels for use by a PBX for voice, a videoconferencing system, and data traffic.

Needless to say, satellite service, LMDS, and MMDS are also potential candidates for transport; they have already been discussed in detail in the prior section, but should be considered if the opportunity to use them arises.

HIGH-BANDWIDTH WIRELESS SERVICES

High-bandwidth point-to-point services fall into two main categories: wireless and wired. Wireless services include microwave, which can provide connectivity at speeds in the Gbps range and infrared, which typically operates in the tens-of-Mbps range. The advantages of both include high transmission speed and relatively low deployment cost. There are also disadvantages. Both require line-of-sight transmission, which means that each end of the circuit must have an unimpeded view of the other end. In microwave transmission, the signal can be repeated (a typical microwave hop is about 30 miles), but this adds cost and complexity to the deployed circuit. Furthermore, microwave is susceptible to atmospheric interference from fog and rain (often called *rain fade*), which lowers its effectiveness in certain geographical areas. The technology also requires an operational license because of the frequency in which it operates and must be monitored carefully to avoid interference with other radio-based services.

Infrared is used for shorter hops because it is extremely secure, operates at a high transmission speed, and requires no operating license. It can be used both inside and outside a building; it is often deployed in secure facilities because it will not propagate outside of the building. (Some radio-based systems will emit energy beyond the confines of the building in which they are deployed, making their use less than desirable when sensitive information is being transmitted.)

Infrared generally does not offer the distance nor the bandwidth that microwave can provide, but it is less expensive. In the local area environment, infrared can easily operate at 100 Mbps with extremely low error rates. In the wide area, it can operate up to 34 Mbps, but often provides lower throughout because of atmospheric interference.

HIGH-BANDWIDTH WIRED SERVICES

In the copper realm, DS-1 (usually called T-1) and DS-3 are the most commonly deployed services for high-bandwidth applications. T-1 operates at 1.544 Mbps, while DS-3 is the equivalent of 28 DS-1s, offering a respectable 44.736 Mbps. Both can be sold as either a channelized or unchannelized service as can their international counterparts, E-1 (2.048 Mbps) and E-3 (usually 34 Mbps).

In North America, the standard digital multiplexing hierarchy that begins with DS-0 effectively ends at DS-3. There have always been multiplexers available operating at speeds in excess of DS-3's 45 Mbps, but until the advent of the SONET, those devices used proprietary framing schemes, which limited their usefulness due to a lack of interoperability with different vendors' devices.

OPTICAL SYSTEMS

Optical transmission has a number of significant advantages over copper-based media. First, it is immune to the electromagnetic interference that plagues metallic media and can therefore be installed with less concern for its proximity to noise-generating devices (electric motors, fluorescent lights, and so on). It is also difficult—although not impossible—to tap and is therefore more secure than copper facilities. It does not require regeneration equipment placed as frequently in the circuit as metallic circuits do because optical signals do not weaken as quickly as electrical signals do. Finally, optical technology provides unprecedented bandwidth—Tbps some cases.

OPTICAL NETWORKING

At their most basic level, optical networks require three fundamental components: a source of light, a medium over which to transport it, and a receiver for the light. Additionally, there may be regenerators, optical amplifiers, and other pieces of equipment in the circuit. We will examine each of these generic components in turn. They are shown in Figure 1-32.

OPTICAL SOURCES   Today the most common sources of light for optical systems are either light-emitting diodes or laser diodes. Both are commonly used, although laser diodes have become more common for high-speed data applications because of their coherent signal. Although lasers have gone through several iterations over the years, including ruby rod and helium-neon, semiconductor lasers became the norm shortly after their introduction in the early 1960s because of their low cost and high stability.

LIGHT-EMITTING DIODES (LEDs)   LEDs come in two varieties: *surface-emitting LEDs* and *edge-emitting LEDs*. Surface-

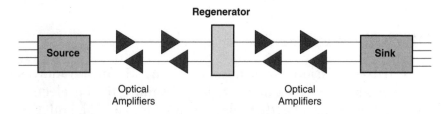

FIGURE 1-32   Optical network components.

emitting LEDs give off light at a wide angle and therefore do not lend themselves to the more coherent requirements of optical data systems because of the difficulty involved in focusing their emitted light into the core of the receiving fiber. Instead, they are often used as indicators and signaling devices. They are, however, quite inexpensive and are therefore commonly found.

An alternative to the surface-emitting LED is the edge-emitting device. Edge emitters produce light at significantly narrower angles and have a smaller emitting area, which means that more of their emitted light can be focused into the core. They are typically faster devices than surface emitters, but do have a downside: They are temperature-sensitive and must therefore be installed in environmentally controlled devices to ensure the stability of the transmitted signal.

LASER DIODES   Laser diodes represent an alternative to LEDs. A laser diode has a very small emitting surface, usually no larger than a few microns in diameter, which means that a great deal of the emitted light can be directed into the fiber. Because they represent a coherent source the emission angle of a laser diode is extremely narrow. It is the fastest of the three devices.

There are many different types of laser diodes. The most common are the *electro-absorptive modulated laser* (EML), which combines a continuous wave laser with a modulating shutter device; the distributed feedback laser, which has an integrated grating assembly to maintain a constant output frequency; and a *vertical cavity surface-emitting laser* (VCSEL, pronounced vick-sel), which produces light from a round spot, resulting in a beam of light that is less prone to spread than a typical surface-emitting laser's output. VCSELs are low-power, low-cost, multifrequency devices. Finally, Fabry-Perot lasers are older devices that suffer a number of problems and are less commonly used. They tend to emit light at multiple closely spaced wavelengths and are commonly called *multimode lasers*.

The surface-emitting LED has the widest emission pattern, followed by the edge emitter; the laser diode represents the most coherent and therefore effective light generator. In fact, the graph of the output signal of an LED versus that of a laser is rather dramatic, as shown in Figure 1-33 (vertical axis not to scale).

OPTICAL FIBER   Generally speaking, there are two major types of fiber: *multimode*, which is the earliest form of optical fiber and is characterized by a large diameter central core, short-distance capabilities, and low bandwidth, and *single mode*, which has a narrow core and is capable of greater distance and higher bandwidth.

Optical fiber has a number of advantages over copper: It is lightweight, has enormous bandwidth potential, has significantly higher tensile strength, can support many simultaneous channels, and is immune to electromagnetic interference. It does, however, suffer from several disruptive problems that cannot be discounted. The first of these is *loss* or *attenuation,* the inevitable weakening of the transmitted signal over distance that has a direct analog in the copper world. Attenuation is typically the result of two subproperties, *scattering* and *absorption,*

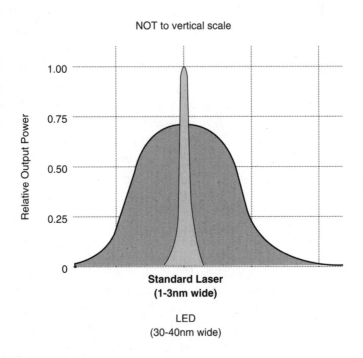

FIGURE 1-33   Emitters.

both of which have cumulative effects. The second is *dispersion*, which is the spreading of the transmitted signal and is analogous to noise.

OPTICAL RECEIVERS (PHOTODETECTORS)   So far, we have discussed the sources of light, including LEDs and laser diodes; we have briefly described the various flavors of optical fiber and the problems they encounter as transmission media; now we turn our attention to the devices that receive the transmitted signal.

The receive devices used in optical networks have a single responsibility: to capture the transmitted optical signal and convert it into an electrical signal that can then be processed by the end equipment. There may also be various stages of amplification to ensure that the signal is strong enough to be acted upon and some demodulation circuitry, which recreates the originally transmitted electronic signal.

PHOTODETECTOR TYPES   Although there are many different types of photosensitive devices, there are two used most commonly as photodetectors in modern networks: *positive-intrinsic-negative* (PIN) photodiodes and *avalanche photodiodes* (APDs).

PIN PHOTODIODES   PIN photodiodes are similar to the device described previously in the general discussion of photosensitive semiconductors. Reverse biasing the junction region of the device prevents current flow until light at a specific wavelength strikes the substance, creating electron-hole pairs and enabling a current to flow across the three-layer interface in proportion to the intensity of the incident light. Although they are not the most sensitive devices available for the purpose of photodetection, they are perfectly adequate for the requirements of most optical systems. In cases where they are not considered sensitive enough for high-performance systems, they can be coupled with a preamplifier to increase the overall sensitivity.

APDS   APDs work as optical signal amplifiers. They use a strong electric field to perform what is known as *avalanche multiplication*. In an APD, the electric field causes current accelerations such that the atoms in the semiconductor matrix get excited and create, in effect, an avalanche of current to occur. The good news is that the amplification effect can be as much as 30 to 100 times the original signal; the bad news is that the effect is not altogether linear and can create noise. APDs are sensitive to temperature and require a significant voltage to operate them—30 to 300 volts depending on the device. However, they are popular for broadband systems and work well in the Gb range.

OPTICAL FIBER TYPES

Fiber has evolved over the years in a variety of ways to accommodate both the changing requirements of the customer community and the technological challenges that emerged as the demand for bandwidth climbed precipitously. These changes came in various forms of fiber that presented different behavior characteristics to the market.

MULTIMODE FIBER

The first of these was *multimode fiber*, which came in a variety of different forms. Multimode fiber bears that name because it enables more than a single mode or ray of light to be carried through the fiber simultaneously because of the relatively wide core diameter that characterizes the fiber (see Figure 1-34). Although the dispersion that potentially results from this phenomenon can be a problem, there were advantages to the use of multimode fiber. For one thing, it is far easier to couple the relatively wide and forgiving end of a multimode fiber to a light source than that of the much narrower single-mode fiber. It is also significantly less expensive and relies on LEDs and inexpensive receivers rather than the more expensive laser diodes and ultra-sensitive receiver devices. However, advancements in technology have caused the use of multimode fiber to fall out of favor; single-mode is far more commonly used today.

Multimode fiber is manufactured in two forms: *step-index fiber* and *graded-index fiber.* Graded-index fiber was commonly

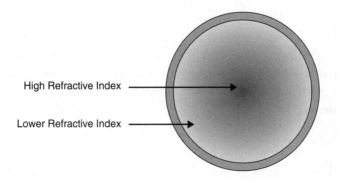

High Refractive Index

Lower Refractive Index

FIGURE 1-34   Multimode fiber.

used in telecommunications applications until the late 1980s. Even though graded-index fiber is significantly better than step-index fiber, it is still multimode fiber and does not eliminate the problems inherent in being multimode. Thus was born the next generation of optical fiber: single-mode.

SINGLE-MODE FIBER

There is an interesting mental conundrum that crops up with the introduction of single-mode fiber. The core of single-mode fiber is significantly narrower than the core of multimode fiber. Because it is narrower, it would seem that its capability to carry information would be reduced due to a limited light-gathering capability. This, of course, is not the case. As its name implies, it enables a single mode or ray of light to propagate down the fiber core, thus eliminating the intermodal dispersion problems that plague multimode fibers. In reality, single-mode fiber is a stepped-index design because the core's refractive index is slightly higher than that of the cladding. It has become the de facto standard for optical transmission systems and takes on many forms, depending on the specific application within which it will be used.

Most single-mode fiber has an extremely narrow core diameter on the order of 7 to 9 microns and a cladding diameter of 125 microns. The advantage of this design is that it only enables a single mode to propagate; the downside, however, is the difficulty involved in working with it. The core must be coupled directly to the light source and the receiver in order to make the system as effective as possible; given that the core is approximately one-sixth the diameter of a human hair, the mechanical process through which this coupling takes place becomes Herculean.

SONET

SONET is an internationally recognized multiplexing standard that begins where DS-3 leaves off. Proposed initially in 1984, SONET was proposed as a solution to the problem of vendor interoperability following the divestiture of AT&T. Integral to the breakup was the concept of equal access, designed to ensure that customers had the right to choose their long-distance carrier from among (at the time) the big three— AT&T, MCI, and Sprint. The problem was that most central offices in the country were replete with Western Electric (aka AT&T) equipment, which meant that if MCI or Sprint wanted to interconnect with the customers (and therefore equipment) served out of a former AT&T central office, they were obligated to purchase Western Electric hardware to ensure interoperability because there were no optical interconnect standards at the time. Clearly, this did not sit well with MCI and Sprint, who did not want to be obligated to one vendor or another, hence the arrival and rapid success of the SONET standard.

SONET's lowest transport speed, called *Optical Carrier Level One* (OC-1) is 51.84 Mbps. Higher speeds can be accommodated by allocating even multiples of OC-1 in a series of recognized combinations, as shown in Table 1-3.

In the same way that T-Carrier systems define both a framing standard (T-1) and a multiplexing hierarchy (DS-1), SONET defines both an *optical carrier level* (OC) and its electrical equivalent, known as the *synchronous transport signal* (STS). Unlike T-Carrier, the OC/STS levels are exact multiples of the base rate. For example, the bandwidth of a T-1 (1.544

TABLE 1-3   Muliples of OC-1

OPTICAL CARRIER LEVEL	ELECTRICAL LEVEL	BANDWIDTH	SDH-STM EQUIVALENT
OC-1	STS-1	51.84 Mbps	—
OC-3	STS-3	155.52 Mbps	STM-1
OC-9	STS-9	466.560 Mbps	STM-3
OC-12	STS-12	622.08 Mbps	STM-4
OC-18	STS-18	933.120 Mbps	STM-6
OC-24	STS-24	1244.16 Mbps	STM-8
OC-36	STS-36	1866.24 Mbps	STM-13
OC-48	STS-48	2488.32 Mbps	STM-16
OC-96	STS-96	4976.64 Mbps	STM-32
OC-192	STS-192	9953.28 Mbps	STM-64

Mbps) is not equal to 24- to 64-Kbps channels because of the added overhead for framing. Similarly, a DS-3's bandwidth is significantly higher than that of 28 DS-1s for the same reasons. In SONET, however, an OC-12 is *exactly* 12 times the bandwidth of an OC-1. This enables SONET (or SDH) to easily and transparently carry payloads at any conceivable speed. A DS-3 fits rather nicely in an OC-1, with a little room left over. A 100-Mbps Fast Ethernet stream can easily be accommodated within a SONET OC-3, with a little left over. SONET operates under the belief that bandwidth is cheap, and it's a good thing because it wastes an awful lot of it in unused bytes.

Like T-Carrier, SONET can be configured as either a channelized or unchannelized service. If it is channelized, multiple OC-1s are multiplexed together on a common fiber. For example, an OC-12 requires 622.08 Mbps of bandwidth, but the fastest speed attainable from the service is OC-1. *The user could derive 12 of them, but they would all operate at the fundamental OC-1 rate.* T-Carrier works in the same way: The user can get 24 64-Kbps channels or a single pipe operating at 1.536 Mbps.

If more bandwidth is required for a higher speed application, the channels can be combined. In this case, our OC-12 becomes an OC-12c, where the c stands for concatenated, which means chained together. Here, we create a service that is analogous to an unchannelized T-Carrier, where the user is provisioned a single circuit that operates at the aggregate rate of the entire facility.

SONET is also fully capable of transporting smaller payloads (that is, payloads that require less bandwidth than an OC-1 provides, such as T-1). In this case, the OC-1 is broken into smaller payload components called *virtual tributaries*, each of which is capable of transporting subrate services. There are four identified virtual tributary types, shown in Table 1-4 with the payload they are each capable of conveniently transporting.

OTHER SONET ADVANTAGES

Besides providing interoperability (often referred to as midspan meet, referring to the capability of two multiplexers from different manufacturers to meet in the middle of an optical span and transparently [no pun intended] swap data), SONET offers a number of other advantages as well. First, it provides a standard multiplexing scheme for services that require bandwidth in excess of DS-3. Second, it provides embedded, very well-designed NM capabilities. Third, it dramatically simplifies the process of adding and dropping (groom and fill) payload com-

TABLE 1-4   Virtual Tributory Types

VIRTUAL TRIBUTARY TYPE	PAYLOAD TYPE	BANDWIDTH
VT1.5	DS-1	1.728 Mbps
VT2	E-1	2.304 Mbps
VT3	DS-1C	3.456 Mbps
VT6	DS-2	6.912 Mbps

ponents along the path. Fourth, it facilitates multipoint configurations, enabling payload to be added and dropped at any point in the network with a minimal amount of additional complexity. Finally, SONET provides for enormous bandwidth, making it possible to transport essentially any payload presented to it.

SONET ARCHITECTURES

SONET networks are usually deployed in dual ring configurations as shown in Figure 1-35. In other words, all elements along the span are connected to dual counter-rotating optical rings, one serving as the primary path, the other as a backup. In this case, the elements on the ring are add-drop multiplexers that have the capability to add and drop payload easily. They are simultaneously connected to both the primary and secondary rings. In the event that the primary ring fails, the devices along the failed path will detect the failure and wrap as shown in Figure 1-36, sealing the ring within 50 ms and preventing a catastrophic failure of the ring. This capability, known as *automatic protection switching* (APS), is one of the key advantages of the SONET ring architecture. It is routinely deployed in major networks to eliminate the possibility of a total failure of the network, something that traditional private lines cannot accomplish.

WDM

The OC-192 (9.9 Gbps) transmission rate that SONET delivers is staggering, yet in many cases is not enough for today's graphic and video-intensive applications. Before the arrival of the Internet and its World Wide Web application, the idea that we would ever need such enormous amounts of transport capacity was ludicrous. However, Parkinson's Law is proving to be true as much in telecommunications as it is in our personal lives.

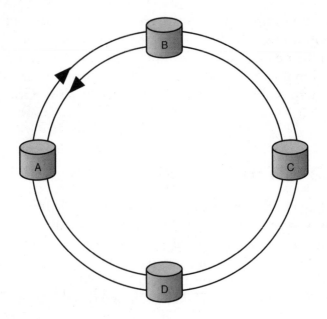

FIGURE 1-35   Optical ring architecture.

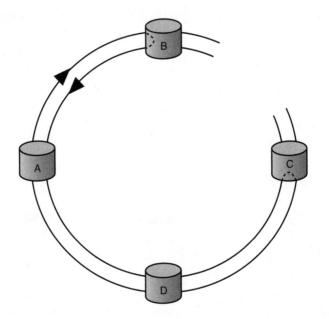

FIGURE 1-36   Failure causes ring to wrap.

The original iteration of Parkinson's Law stated, "Work expands to fill the time allotted to it." A second variation observed, "Expenditures will always rise to exceed income." Today, Parkinson's Law tells us that there is no such thing as a fast enough processor, enough installed memory, a big enough hard drive, or too much bandwidth. Traditional optical fiber is proving to be inadequate for current bandwidth demand. In response, two solutions emerged. The first is to simply deploy more fiber, which solves the problem but at enormous expense —as much as $70,000 per mile in some cases. Alternatively, a relatively new technology has emerged called WDM. Capable of multiplying the overall capacity of a single fiber several times over, WDM lowers the cost of deployed bandwidth from $70,000 per mile to as little as $20,000 because the change requires no backhoe. A simple replacement of endpoint electronics takes care of the conversion.

HOW IT WORKS

In optical systems, light from a high-quality laser (sometimes a LED) is transmitted down the fiber to a receiver, which converts the optical energy to an electrical signal so that it can be processed. The light transmitted is in the infrared range of the electromagnetic spectrum and is therefore not visible. A typical fiber-based system without WDM transmits a single information stream at high speed, typically as high as 2.4 Gbps (OC-48). With WDM technology, that same fiber can carry multiple streams of information at high speeds, thus multiplying the actual capacity of each fiber.

The original WDM systems carried four individual channels, while modern systems can carry as many as 16 or more. Experimentation has shown that it is possible to build systems capable of transporting as many as 1,000 distinct channels, although it will be some time before they become commercially feasible. A number of vendors now sell multichannel systems;

Lucent, Nortel, Ciena, Corvis and Fujitsu, to name a few, have demonstrated high-capacity systems with as many as 160 supported channels. Others will certainly follow with even more densely packed channels.

Among the ranks of the so-called bandwidth barons, Level 3 and Qwest have made enormous strides with the deployment of their own networks. Level 3, which started life as part of a construction company (the same company that gave birth to Metropolitan Fiber Systems), continues to expand a 15,000-mile fiber network over which it plans to run IP throughout North America. They acquired softswitch manufacturer Xcom, which integrates circuit- and packet-based traffic for IP transport. Qwest, spun off from Southern Pacific Railroad in 1988, has built an 18,500-mile network and is now one of the largest long-distance companies in the United States. The firm also has a significant presence in both Asia and Europe through a series of alliances and acquisitions.

The fiber manufacturers themselves have also done a great deal to increase the distance and carrying capacity that their products support. Corning's LEAF fiber, introduced in February 1998, does not need regenerators as often as other fibers and requires less power to create a transportable signal. Most fiber is optimized for regeneration every six miles or so; LEAF only requires regenerators every 30 miles. Lucent's TrueWave and Alcatel's TeraLight fiber products are equally innovative and continue to advance.

FIBER TRANSMISSION ISSUES AND CONCERNS

Although optical fiber offers all the advantages cited earlier—immunity from noise and interference, high security, and enormous bandwidth and distance—it does have some disadvantages as well. First, although it is not affected by electromagnetic interference, there are optical interference problems that occur within the fiber that limit the overall bandwidth of the facility. These include physical limitations of the electronics, distance-dependent signal degradation, *Stimulated Raman Scattering* (SRS), *Stimulated Brouillian Scattering* (SBS), cross-wave modulation, and four wave mixing. We will discuss each of these in turn.

Limitations of the electronics: Simply stated, lasers take a certain amount of time to cycle between different states (the so-called rise time), and the degree to which that period of time can be minimized is finite. There is, therefore, a physical limitation based on solid state electronics as we know them today.

Distance-dependent signal degradation: Even optical fiber suffers from signal degradation due to a number of factors, such as chromatic dispersion, pulse spreading over distance, and a variety of others that will be discussed shortly. It is, therefore, necessary to install regenerators along optical spans that receive the incoming optical signal, convert it to electrical, clean and regenerate the original signal, and convert it back to optical for retransmission. An innovative technology has emerged in recent years that enables optical amplification of the signal (that is, the signal does not have to be converted back to electrical, amplified, and reconverted to optical again), known as the *Erbium Doped Fiber Amplifier* (EDFA). In EDFA systems, a segment of fiber that has had erbium atoms mixed into the silica matrix of the fiber (a process called *doping*) is inserted in the optical strand. The EDFA segment is connected to optical electronics that have the capability to pump energy into selected frequencies—specifically those frequencies used to transport the actual content (between 1480 and 1600 nm). When the signal arrives at the EDFA segment, the incoming light pulse excites the erbium atoms embedded in the EDFA. (Sorry, a little physics is necessary here to illustrate the magic of this technology.) When the erbium atoms get excited, they jump up to the next quantum energy level, and when they fall back down again, they give off a photon whose energy level *happens to fall within the 1480 to 1600 nm range used by optical systems.*[19] The resulting photons pump energy into the passing signal, effectively amplifying it along the way.

SRS is a phenomenon that typically affects large systems and occurs when high-frequency channels in a *dense WDM* (DWDM) system pump energy into low-frequency channels, resulting in high-frequency signal loss. Short wavelength, high-energy channels are the most depleted by this phenomenon, and the only way to avoid it is to operate the system at low power levels. This problem is also a benefit: SRS is precisely the phenomenon used in EDFAs!

SBS affects smaller systems, and only occurs when adjacent channels operate in opposite directions. In SBS, optical energy is scattered back toward the transmitter, resulting in loss of signal power. Again, keeping the system below a predefined power threshold is the only way to eliminate SBS.

Cross phase modulation is a form of crosstalk that is the result of the fact that the refractive index of fiber changes with the intensity of the signal being propagated. As the intensity of the signal changes, so too does the phase shift that naturally occurs during transmission. The phase shift results in crosstalk, which like so many other impairments, can only be eliminated by reducing the overall power of the transmitted signal.

Four wave mixing occurs when the frequency spacing between channels is so small that the different waves phase-lock. This results in the creation of sideband channels that carry no information, but tend to bleed power out of the primary signals. The solution is to use *nonzero dispersion shifted fiber* (NZDSF), which was designed to enable a small amount of signal dispersion without the zero-dispersion-point falling within the WDM passband. Examples of NZDSF include Lucent's Truewave, Corning's LS Fiber, and Corning's LEAF fiber.

---

[19]This range was not chosen randomly. At 1480 and 1550 nm, optical signals are least affected by the various physical impairments that weaken or disperse them, such as hydroxyl (water) absorption.

A WORD ABOUT OPTICAL SWITCHING

The principal form of optical switching is really nothing more than a sophisticated digital cross-connect. In the early days of data networking, dedicated facilities were created by manually patching the end points of a circuit at a patch panel, thus creating a complete four-wire circuit. Beginning in the 1980s, digital cross-connect devices, such as AT&T's *digital access and cross-connect system* (DACS), became common, replacing the time-consuming, expensive, and error-prone manual process. The digital cross-connect is really a simple switch, designed to establish long-term temporary circuits quickly, accurately, and inexpensively.

Enter the world of optical networking. Traditional cross-connect systems worked fine in the optical domain, provided there was no problem going through the *optical-electrical-optical* (O-E-O) conversion process. This, however, was one of the aspects of optical networking that network designers wanted to eradicate from their functional requirements. Thus was born the optical cross-connect switch.

The first of these to arrive on the scene was Lucent Technologies' LambdaRouter. Based on a switching technology called *micro electrical mechanical system* (MEMS), the LambdaRouter was the world's first all-optical cross-connect device.

MEMS relies on micro-mirrors, an array of which is shown in Figure 1-37. The mirrors can be configured at various angles to ensure that an incoming lambda strikes one mirror, reflects off a fixed mirrored surface, strikes another movable mirror, and is then reflected out an egress fiber. The LambdaRouter is now commercially deployed and offers speed, a relatively small footprint, bit rate and protocol transparency, non-blocking architecture, and highly developed database management.

The mirror-based MEMS technology is the best-known wavelength switching technique, but other technological contenders have recently entered the marketplace and are showing

FIGURE 1-37   MEMS array (Courtesy Lucent Technologies).

great promise. The Agilent Photonic Switching Platform includes a 32×32 port and a dual 16×32 port photonic switch, both of which include switch control and management electronics and a well-designed *application programming interface* (API). Based on a combination of the world-famous Agilent inkjet technology and a relatively new technology called *planar lightwave circuits*, the switch is capable of moving optical signals without the moving parts found in MEMS-based devices. The switch comprises an array of vertical and horizontal waveguides that are permanently aligned. Optical signals are transmitted across the horizontal waveguides from an input port to an output port. When a switch command is received, a bubble is formed by applying heat at the intersection of the horizontal path along which the signal is traveling and the vertical path that leads to the new output port.

Alcatel Optics and Agilent have joined forces to develop switching elements based on this new technology. This relationship will lead to the development of optical cross-connect switches, optical add-drop multiplexers, and optical-based protection switching equipment.

OTHER SWITCHING SOLUTIONS

Manufacturer Gooch and Housego PLC have developed acousto-optical switches that actually use sound waves to reliably deflect light from one fiber to another. Using a fused coupler that attaches two fibers to each other, specific wavelengths of light are forced from one fiber to the other. These devices have no moving parts and suffer very little loss compared to MEMS-based devices. They do have a downside, however: They are slow and can be expensive.

SONET VERSUS DWDM: A TECHNOLOGY BATTLE

There is a battle underway between those in the SONET camp and those in the DWDM camp over the underlying nature of the future transport network, and it is worthwhile to present both sides of the argument here. The argument centers on whether future networks will transport their payloads (probably IP packets) within the SONET framing structure or will simply hand them directly down to DWDM for transmission.

SONET is a physical layer protocol that defines a multiplexing scheme for the transport of a wide variety of services over a common optical network. It includes a well-defined NM system as well as a highly developed self-healing architecture to prevent catastrophic failures in the event of a span disruption. In the event that a ring is cut, the automatic protection switching bytes kick in and cause the failed ring to be abandoned in favor of the backup ring, thus preventing interruption of service.

DWDM, on the other hand, is a highly effective multiplexing scheme that is not yet widely standardized and that does not yet offer any sort of survivability guarantees along the lines of SONET's APS capability. The *Optical Internetworking Forum* (OIF) was created to develop a strategy for the widespread deployment of DWDM architectures and to guide the development of globally accepted standards that will ensure interoperability.

Today, it is common to see the following architecture in high-speed networks: IP packets are carried within ATM cells, which are in turn transported within SONET frames over a DWDM-enhanced, dual counter-rotating ring-based fiber network. This architecture is highly survivable and makes the adding and dropping of payload components extremely easy. It is also largely standards-based, because IP, ATM, and SONET are all based upon widely accepted international standards blessed by the IETF, the ATM Forum, and the ITU, to name a few. Meanwhile, the OIF has focused its energies on three critical areas for DWDM: circuit protection and restoration, integrated NM systems similar to those provided by SONET, and interoperability among optical interface equipment.

Although DWDM is perfectly capable of transporting IP packets at blindingly fast speeds, it does not yet have the capability to guarantee survivability, to provide interoperability, or to offer NM, and although ATM and SONET add a significant amount of overhead (and therefore inefficiency) to the transported signal, most believe that it is a fair price to pay for the peace of mind that comes from an overdesigned network that is not likely to fail. Because the majority of those looking to deploy DWDM-based fiber networks are the large, incumbent carriers whose primary focus has always been on customer service, they are not likely to surrender their ATM/SONET-based networks in the near-term. However, with work underway to augment the DWDM technology to include interoperability and survivability guarantees, it is possible that both SONET and ATM could become unnecessary in the future.

Some companies have already experimented with alternative architectures. In May 1999, Cambrian Systems combined their efforts with those of 3Com, Bay Networks, and Packet Engines to prove that SONET was not required for survivable transmission. Instead of ATM at the switching layer, they used gigabit Ethernet and created a highly efficient system that did not require the massive overhead that SONET brings with it. Today, Ethernet has found a new home as a low-cost, high-speed transport scheme, and is in fact the primary deliverable of such companies as Telseon and Yipes. Widespread implementation, however, is a long way off; traditional, widely accepted architectures will remain in use for some time to come. It was the telephone company, after all, that coined the phrase, "If it ain't broke, don't fix it."

We will examine the SONET/DWDM/Ethernet debate in greater detail in a later chapter when we discuss QoS issues.

POINT-TO-POINT
TECHNOLOGIES SUMMARY

Point-to-point technologies range from very low-speed, analog modem-based systems for telemetry and alarm circuit applications to unimaginably high-bandwidth services that operate in the multiple Gbps range. For implementations that require the capability to transfer information between two nonchanging points or between a small number of points, point-to-point solutions represent a reasonable solution. They are secure, are universally available, offer minimal traffic delay, and are available with tremendous bandwidth granularity. They do have a number of downsides, however, that must be taken into account when selecting network architectures. First of all, they are dedicated, which means that once they are established, they cannot be redirected to another site without a service order. Second, they are generally billed based on bandwidth and circuit miles, which means that the price does not change from month to month—there is no usage component. The circuit costs the same whether it is in use 2 percent of the time or 100 percent of the time, which means that it is most economical when it is in full-time use.

Although these downsides may appear trivial in light of the relative advantages of dedicated facilities, they can become major issues when flexibility, distance, and in some cases bandwidth concerns come into play. The result has been a migration to switched technologies that overcome all of these disadvantages while at the same time offering services that emulate a private line.

SWITCHED TECHNOLOGIES

Switched technologies came about because of the high cost and inflexibility characteristic of early dedicated communications architectures. When telephony was first introduced, there was no concept of switching. There were so few customers that it was common practice to run dedicated telephone circuits between individuals who wanted to call one another. As telephones became more common and call volumes increased, the need for switching arose. The first switches were people who asked the customer for the number they wanted to reach, and then inserted a patch cord that interconnected the caller with the called party. In the early days, when call volumes were still relatively low, the technique of using human operators was adequate to handle the number of incoming call setup requests. However, as call volumes increased, human operators could no longer handle the task, and mechanical switches were invented.

The first mechanical switch was invented in 1891 by Almon Strowger, a Kansas City undertaker. According to local legend, Strowger realized one day that his business volume had dropped off, and because the average death rate in Kansas City had not declined, he became suspicious. Upon investigation he learned that the town's telephone operator was married to Strowger's competitor, which meant that when calls for an undertaker came in, they were not routed to Strowger. In response, he designed and built the first mechanical switch, called a *step-by-step*, which when installed gave customers the ability to make their own routing decisions by dialing numbers instead of depending on the operator to do it for them.

As time went on, mechanical step-by-step switches were replaced by various generations of faster, more efficient crossbar switches, which in turn were replaced by several succeeding generations of electronic switches. Modern central office switches, such as the Lucent Technologies 5ESS, the Nortel Networks DMS100, the Siemens EWSD, and the Ericsson AXE, are all examples of high-speed, high-volume devices capable of processing more than 100,000 calls per hour using a technique called *circuit switching*.

CIRCUIT SWITCHING

The telephone network is a perfect example of circuit-switching. When a customer wants to place a call, he picks up the handset, which notifies the switch of his desire to do so, and the switch responds by sending a dial tone to the phone. The caller then sends the destination address to the switch (the telephone number), which proceeds to establish the call. As soon as the other end goes off-hook, the call is established.

The primary characteristic of this technology is its dedicated nature. For the duration of the call, all of the loop, switch, and trunk resources necessary to establish the call are dedicated to the two endpoints and cannot be shared with other users. As long as the call is in progress, the resources are off-limits to the network. As a consequence of this, circuit-switched networks must be massively overdesigned to accommodate the high volume of calls that they experience, particularly on holidays like Mother's Day. Circuit-switched networks are, therefore, expensive and inefficient but extremely capable. The service they provide emulates the service delivered by a private-line circuit, albeit typically with less bandwidth than a high-end private-line circuit can provide. They work well for such applications as voice, low-speed video, medium-speed data, and dial backup, and the technology is therefore deeply embedded in the network.

Circuit switching has experienced a major problem in recent years as Internet access with its attendant Web surfing has grown popular. Because call resources are dedicated, the use of circuit-switched facilities for modem access to the Web has resulted in the switches becoming severely overtaxed by the long call-hold times that characterize most Web sessions. When it becomes widely deployed, DSL will do a lot to eliminate the extra load on the switched network, but in the long run, packet switching represents the best solution for data transport. Circuit switching is still the best solution for voice traffic, particularly if the data packets can be routed elsewhere. Enter packet switching.

PACKET SWITCHING

Packet switching has its roots in a technique known as *store-and-forward switching*, which was first implemented as a technology called *message switching*. Store-and-forward systems do precisely what their name implies. Upon receiving information to be transported, the switch stores it in some form of memory, checks it for errors, makes a routing decision, and forwards the message out the proper port to the next switch.

Message switching is not used much anymore, but understanding how it worked helps to understand the mechanics of modern switch fabrics. In message switching, information to be transported across the network was sent from switch to switch as an entire message, a technique that had several downsides. First, it required the switches to have hard drives because messages were too large to store in the memory available in switches at the time. This meant that the process was slow because of the need to involve a mechanical device to process the message. If a message arrived at a switch and upon examination was found to have errors, it was discarded, and a second copy was requested from the prior switch. This is clearly wasteful; a single bit error could result in the need to retransmit an entire message. Early teletype systems relied on message switching, but the technology has largely been abandoned in favor of packet switching.

Packet switching is a data transmission technique in which messages (which can be voice, video, images, or pure data) are broken into small segments called *packets*. Each packet is then encapsulated within a frame that may contain information about the nature of the data being transported, its relative level of criticality, its position within the data stream (one of five, two of five, three of five, and so on), some error correction information, and addressing overhead.

Unlike their message-switched predecessors, packet-switched networks are quite efficient. They take into account the fact that data, as a general rule, is bursty, meaning that typical data users do not send continuous streams of information but rather somewhat random bursts of traffic. As a consequence, they do not use the transmission facility 100 percent of the time. It may be idle for considerable periods of time. To overcome this inherent inefficiency, packet networks combine the packet streams from multiple users onto a single facility, thus using a higher percentage of the available bandwidth. Control protocols manage the independent data streams to ensure that each user data component maintains its integrity. In some cases, this technique is called *virtual circuit service*; the name stems from the fact that the service is so good in these networks that users feel as if they have their own dedicated transmission channel, in spite of the fact that the channel is shared among a collection of users. The word virtual becomes rather central in data communications because it defines a family of quite lucrative services that don't really exist! Within the confines of communications, the word virtual has only one meaning: *It's a lie.* If you see the word used to describe a technology or service, you should immediately think *they're lying to me. It isn't real.* Yet these services are extremely capable, cost effective, and popular, as we will see.

CONNECTIONLESS PACKET-SWITCHED SERVICE

Packet networks provide two forms of service: connectionless service, where the switches do not perceive a relationship between packets that derive from the same source, and connection-oriented service, where they do. In connectionless networks, each packet contains complete destination address information, and as a result, every packet can be routed independently of the others. The benefit of this is that connectionless networks are extremely survivable because packets are not required to take any particular path through the network fabric and can be rerouted as the network sees fit. The disadvantage is that because of this seemingly random routing mechanism, packets can arrive out of order, resulting in the need for a higher-layer protocol to put them back into the correct order. Connectionless network protocols include IP among others, and the service they deliver is called *datagram*.

Connectionless networks make no guarantees of delivery or arrival sequence. They are often called *best effort* or *spray and pray* networks because although they make every effort to deliver the packets to the destination, they do not guarantee the delivery. That is the responsibility of a higher-layer protocol. For example, IP depends on TCP for ironclad, guaranteed delivery of transmitted packets if such guarantees are required. Their primary advantage comes from their inherent capability to route around trouble areas in the network, thus ensuring survivability.[20]

CONNECTION-ORIENTED PACKET-SWITCHED SERVICE

Connection-oriented networks differ from their connectionless counterparts in that they establish a path based on known network conditions and send all packets from the same source

---

[20]The Internet, which is connectionless, was originally conceived as a Department of Defense communications network during the 1960s, when the fear of nuclear attack was in the forefront of everyone's mind. IP was chosen because it would enable a network to continue operation even if one or more nodes were lost due to attack.

across the same path, thus ensuring sequential delivery. Because this service emulates private line, it is often called *virtual circuit service*.

In connection-oriented networks, there is a multistage process that must be adhered to before information is transmitted across the network. In stage one, a call setup packet (I like to call it a Lewis and Clark packet) is sent into the network. This initial packet contains the full address of the intended recipient of the message. Upon arrival at the first switch, the packet is examined, the address is read, and a route (outgoing port) is selected based upon known information about the network and the intended destination. The switch then makes an entry into its routing table indicating which port the packet arrived on and which one it went out through. The packet continues across the network, causing table entries to be written at each switch, and thus blazes a trail across the network. Once the trail has been laid down, the remaining packets can make their way across the digital continent, following the blazes (table entries) left on the trees (switches) that they encounter along the way. As a result, the packets that follow the setup packet do not require a full address. All they need is a short virtual circuit identifier; the switches do the rest.

The advantage of connection-oriented packet switching is that unlike connectionless service, all packets take the same route through the network, thus ensuring sequential delivery. The technology emulates the service a customer would receive from a dedicated facility. Examples of connection-oriented technologies include X.25 packet switching, frame relay, and ATM.

A word about packet technology: Once again, remember that this game is not about technology, but rather about providing customer service. The primary beneficiary of packet-switched technology is not the customer. If the service provider builds a robust, properly engineered packet-switched network that is capable of transporting a wide variety of service types throughout their operating area, and if they implement QoS protocols such that they can guarantee the services they sell to the customer, then they are in a position to be all things to all applications. If they then combine their advanced network fabric with a wide variety of access offerings, customers will have no reason to go anywhere else.

Remember the cloud? If service providers design the right cloud, they can satisfy every possible user demand for service, and customers will be happy to buy their gozintas and gozoutas from the service provider. If the customer wants to buy a private-line service that operates at a particular speed, the service provider can deploy the circuit over a switched fabric that is *shared* but which gives the appearance of being *dedicated*. By taking advantage of the inherently efficient nature of packet technology, the service provider can build a far more cost-effective network while continuing to satisfy customer demands for high-quality, dependable service. Ultimately, the customer doesn't care and doesn't have to know. All they care about is that they are getting what they asked for, and the underlying technology, although critically important, remains invisible to them.

In order for this concept to work, a newer form of packet technology must be introduced. Datagram and traditional virtual circuit services are fine for the limited requirements of low and medium-speed data—the services they were originally designed for—but they fall short of being able to provide for modern applications that require greater bandwidth allocations. For those services, *fast packet technology* is required.

FAST PACKET SERVICES

Fast packet technologies include such services as frame relay and ATM and are considered fast for several reasons. The principal reason is that they have the capability to dramatically reduce the overhead required to receive, examine, and forward a packet of data by as much as 80 percent.

In order to achieve these remarkable improvements, fast packet technologies make a number of assumptions about the network. First, they assume that it is based upon a fiber infrastructure, from which they can further assume a low error rate. Second, they assume modern, robust switch fabrics, which also offer low error rates. Finally, they make the assumption that the end devices possess a modicum of intelligence and can thus make informed decisions about the correctness of the received data.

When a traditional packet switch receives a packet, it goes through the following steps. First, it buffers the packet. Next, it invokes an error detection algorithm and checks the packet for bit errors. If it finds one, it invokes an error *correction* algorithm, which in most cases causes the switch to go back to the last switch in the chain and ask for another copy of the packet, while at the same time discarding the bad one. Once it receives a copy that it determines to be error-free, the switch goes to its routing table, selects a route for the packet, and transmits it to the next switch in the chain.

In fast packet environments, the process is dramatically simplified. The switch receives a packet, and rather than buffer the packet, it often implements an option called *cut-through*. With cut-through invoked, the switch never buffers the packet, but rather proceeds to parse it for errors while at the same time examining the destination address information and selecting an outgoing port. In effect, the head of a packet can already be on its way out of the switch while the tail is still being examined for errors. In the event that an error is detected, there is no attempt to correct it. In fact, the switch does not ask for a resend, nor does it attempt to notify anyone of the bad packet. It simply throws the bad packet away, without emotion, and makes the correct assumption that the end stations have enough innate intelligence to detect the error and take whatever steps are necessary to correct it. This process is significantly more efficient than the store-and-forward technique described earlier.

As we mentioned earlier, frame relay and ATM are the most common examples of fast packet switching.

FRAME RELAY

Often described as X.25 on steroids, frame relay came about as a private-line replacement technology and was intended as a data-only service. Today, it routinely carries not only data, but voice and video as well, and although it originally emerged with a top speed of T-1/E-1, it now provides connectivity at much higher speeds. In frame relay networks, the incoming data stream is packaged as a series of variable length frames[21] that work like the telecommunications equivalent of a Trojan horse. A frame relay frame can transport any kind of data—LAN traffic, IP packets, *systems network architecture* (SNA) frames, even voice and video in certain cases. In fact, of late it has been recognized as a highly effective transport mechanism for voice, enabling frame relay-capable PBXs to be connected to a frame relay *Permanent Virtual Circuit* (PVC), which can cost effectively replace private-line circuits used for the same purpose. When voice is carried over frame relay, it is usually compressed and packaged in small frames for transport to minimize the processing delay of the frames. According to the Frame Relay Forum, as many as 255 voice channels can be encoded over a single PVC, although the number is usually smaller when implemented.

Frame relay is a virtual circuit service. When a customer wants to connect two locations using frame relay, they call the service provider and indicate to the service representative the locations of the endpoints and the bandwidth they require for the applications that will operate across the circuit. The service provider issues a service order that results in the establishment of the circuit. If at some point in the future the customer decides to change the circuit endpoints or upgrade the bandwidth, another service order must be issued. This service is called PVC and is the most commonly deployed frame relay offering.

---

[21]In telecommunications parlance, a frame is a distinguishable entity that has a clear beginning and end, a variable-length field for application data, and some form of error detection mechanism. In many cases, the user data contained within a frame is a packet.

Frame relay is also capable of supporting *switched virtual circuit* (SVC) service, but SVCs are not currently available. With an SVC service, customers can make their own modifications to the circuit by accessing the switch and requesting changes. However, service providers do not currently offer the service because of billing and tracking issues. Instead, they allow customers to create a fully meshed network between all their locations that enables any user on the network to send traffic to any other user on the network. Instead of making routing changes in the switch, the customer has a circuit between every possible combination of desired endpoints. The service provider usually makes this a cost-effective option for the user by charging a normal monthly fee for the first PVC and very low fees for each additional PVC. As a result, customers get the functionality of a switched network, while the service provider avoids the difficulty of administering a network within which the customer is actively making changes.

In frame relay, PVCs are identified using an address called a *data link connection identifier,* or DLCI (pronounced delsie). At any given endpoint, the customer's router can support multiple DLCIs, and each DLCI can be assigned varying bandwidths based upon the requirements of the device/application on the router port associated with that DLCI. Let's say a customer has purchased a 768-Kbps circuit to connect his router to the frame relay network. The router is connected to a videoconferencing unit at 384 Kbps, a frame relay-capable PBX at 384 Kbps, and a data circuit for Internet access at 128 Kbps. Note that the aggregate bandwidth assigned to these devices exceeds the actual bandwidth of the access line by 128 Kbps. Under normal circumstances this would not be possible, but frame relay assumes that the traffic that it will normally be transporting is bursty by nature. If the assumption is correct (and it usually is), there is very little likelihood that all three devices will burst at the same instant in time. As a consequence, the circuit's operating capacity can actually be overbooked, a process known as *oversubscription.* Most service providers allow as much as 200 percent oversubscription, something customers clearly benefit from provided the circuit is designed properly.

This means that the salesperson must carefully assess the nature of the traffic that the customer will be sending over the link and ensure that enough bandwidth is allocated to support the requirements of the various devices that will be sharing access to the link. Failure to do so can result in an underengineered facility that will not meet the customer's throughput requirements. *This is a critical component of the service delivery formula.*

The throughput level, that is, the bandwidth that frame relay service providers absolutely guarantee on a PVC-by-PVC basis, is called the *committed information rate* (CIR). In addition to CIR, service providers will often support an *excess information rate* (EIR), which is the rate above the CIR they will attempt to carry, assuming the capacity is available within the network. However, all frames above the CIR are marked as eligible for discard, which simply means that the network will do its best to deliver them but makes no guarantees. If push comes to shove and the network finds itself to be congested, the frames marked *discard eligible* (DE) are immediately discarded at their point of ingress. This CIR/EIR relationship is poorly understood by many customers because the CIR is taken to be an indicator of the absolute bandwidth of the circuit. Whereas bandwidth is typically measured in bits per second, CIR is a measure of *bits in one second*. In other words, the CIR is a measure of the average throughput that the network will guarantee. The actual transmission volume of a given CIR may be higher or lower than the CIR at any point in time because of the bursty nature of the data being sent, but in aggregate the network will maintain an average, guaranteed flow volume for each PVC. This is a selling point for frame relay. In most cases, customers get more than they actually pay for, and as long as the switch is properly engineered, it will not suffer adversely from this charitable bandwidth allocation philosophy. The key to success when selling frame relay is a very clear understanding of the applications the customer intends to use across the link so that the facility can be properly sized for the anticipated traffic load.

SERVICE GRANULARITY
IN FRAME RELAY

Frame relay does not offer a great deal of granularity when it comes to QoS. The only inherent mechanism is the discard eligibility bit (DE) described earlier as a way to control network congestion. However, the DE bit is binary; it has two possible values, which means that a customer has two choices: the information being sent is either important, or it isn't—not particularly useful for establishing a variety of QoS levels. Consequently, a number of vendors have implemented proprietary solutions for QoS management. Within their routers (sometimes called *frame relay access devices*, or FRADs), they have established queuing mechanisms that enable customers to create multiple priority levels for differing traffic flows. For example, voice and video, which don't tolerate delay well, could be assigned to a higher priority queue than the one to which asynchronous data traffic would be assigned. This enables frame relay to provide a highly granular service. The downside is that this approach is proprietary, which means that the same vendor's equipment must be used on both ends of the circuit. Given the strong move toward interoperability, this is not an ideal solution because it locks the customer into a single vendor situation.

CONGESTION CONTROL
IN FRAME RELAY

Frame relay has two congestion control mechanisms that should be mentioned. Embedded in the header of each frame relay frame are two additional bits called the *forward explicit congestion notification bit* (FECN) and the *backward explicit congestion notification bit* (BECN). Both are used to notify devices in the network of congestion situations that could affect throughput.

Consider the following scenario. A frame relay frame arrives at the second of three switches along the path to its intended destination, where it encounters severe local congestion. The congested switch sets the FECN bit to indicate the presence of congestion and transmits the frame to the next switch in the chain. When the frame arrives, the receiving switch takes note of the FECN bit, which tells it the following: "I just came from that switch back there, and it's extremely congested. You can transmit stuff back there if you want to, but there's a good chance that anything you send will be discarded, so you might want to wait awhile before transmitting." In other words, the switch has been notified of a congestion condition, to which it may respond by throttling back its output to enable the affected switch time to recover.

On the other hand, the BECN bit is used to flow-control a device that is sending too much information into the network. Consider a situation where a particular device on the network is transmitting at a high volume level, routinely violating the CIR and perhaps the EIR level established by mutual consent. The ingress switch—that is, the first switch the traffic touches—has the capability to set the BECN bit on frames going toward the offending device, which carries the implicit message, "Cut it out or I'm going to hurt you." In effect, the BECN bit notifies the offending device that it is violating protocol, and continuing to do so will result in every frame from that device being discarded—without warning or notification. If this happens, it gives the ingress switch the opportunity to recover. However, it doesn't fix the problem—it merely forestalls the inevitable because sooner or later the intended recipient will realize that frames are missing and will initiate recovery procedures, which will cause resends to occur. However, it may give the affected devices time to recover before the onslaught begins anew.

The problem with FECN and BECN lies in the fact that many devices choose not to implement them, choosing instead to rely on the Sylvester Stallone protocol. Allow me to explain. Many devices, upon receiving a frame with a set FECN or BECN bit, respond with the digital equivalent of (Sly's voice, now), "Yeah. So?" They do not necessarily have the inherent capability to throttle back upon receipt of a congestion indicator, although devices that can are becoming more common.

Nevertheless, proprietary solutions are in widespread use and will continue to be for some time to come.

FRAME RELAY SUMMARY

Frame relay is clearly a Cinderella technology, evolving quickly from a data-only transport scheme to a multiservice technology with diverse capabilities. For data and some voice and video applications, it shines as a WAN offering. In some areas, however, frame relay is lacking. Its bandwidth is limited to DS-3, and its capability to offer standards-based QoS is limited.

ATM

Like frame relay, ATM is a connection-oriented technology that has assumed the mantle of leadership among current WAN offerings for a number of reasons that will be discussed shortly.

Frame relay and ATM share both similarities and differences. Both are virtual circuit-based fast packet services for wide area transport that support a variety of payload types. Whereas frame relay's bandwidth is limited to 45 Mbps, ATM tops out at the upper limits of SONET speed, currently OC-192 (approximately 10 Gbps). Where frame relay requires proprietary solutions for granular QoS, ATM offers an array of internationally accepted, widely standardized QoS definitions that enable it to support a plethora of services with ease, in spite of being characteristically different in terms of relative levels of necessary bandwidth, burstiness exhibited, and QoS required. Furthermore, unlike frame relay, ATM's transmission unit is a fixed-size cell rather than a variable-size frame. This has a number of advantages. First, it simplifies the design of the switch. In frame relay, the switch must be capable of processing frames that arrive in a variety of sizes, which makes buffering and switching a complex process. In cell relay, of which ATM is an example, all cells are exactly the same size (a 5-octet header that contains information the switch uses to properly route the cell and a 48-octet payload field that contains the user's data). Regardless of the payload type, the switch can always expect to receive protocol data units of the same size, which means that its buffering is simplified, and the switch is therefore very fast.

ADDRESSING IN ATM

In ATM, virtual circuits are identified by a combination of *virtual path identifiers* (VPI) and *virtual channel identifiers* (VCI). A single virtual circuit is identified by a VPI/VCI combination. VCIs are unidirectional channels, while a VPI is a combination of VCIs that results in a bidirectional circuit. When a device wants to send traffic to another device in an ATM network, it addresses the message to the appropriate virtual path, which by definition identifies the virtual channels.

QoS AND CLASS OF SERVICE (CoS) CONTROL IN ATM

The terms QoS and CoS are used interchangeably in discussions about transport technologies. They are, however, quite different, and it would be useful to make the distinction here.

QoS is a measure that is inviolable. It is absolute in the sense that when QoS is implemented in a network the provider guarantees that they will deliver measurable performance characteristics as defined by the QoS statements in the *service level agreement* (SLA) that exists between the provider and the consumer of the service.

CoS, on the other hand, is defined by the provider of the service and is therefore a relative measure, unrelated to measures of quality. For example, a service provider would be perfectly justified (although not particularly bright) if they were to identify and advertise the following transmission classes of service:

- Really good
- Pretty good
- Not so great

- Fair
- So-so
- Better write a letter

More reasonable classes might define services that align accurately with the types of application traffic transported by the network provider for the customer.

ATM has experienced several generations of service class definitions over the years, including efforts by both the ITU and the ATM Forum. By and large, all of the efforts have revolved around three service characteristics: whether the traffic's bit rate is constant or variable, what kind of timing relationship exists between the end points of the facility, and whether the circuit is connectionless or connection-oriented. Using these three characteristics, the ITU developed the model shown in Figure 1-38. Known as the *Broadband ISDN protocol model*, it enables service providers to define classes of service that meet the transport requirements of every possible traffic type. Broadband ISDN has little to do with traditional narrowband ISDN other than the fact that it defines a broadband digital network over which a set of integrated services can be transmitted.

	Class A	Class B	Class C	Class D
AAL Type	1	2	5, 3/4	5, 3/4
Connection Mode	Connection-oriented	Connection-oriented	Connection-oriented	Connectionless
Bit Rate	Constant	Variable	Variable	Variable
Timing Relationship	Required	Required	Not required	Not required
Service Types	Voice, video	VBR voice, video	Frame relay	IP

FIGURE 1-38   The ATM service model.

Over time and as a result of a shared effort between both the ITU and the ATM Forum, the parameters for assigning service definitions (QoS) evolved. Similar to frame relay, ATM now relies on cell rate indicators to identify service flows within the network.

To understand QoS in ATM, it is useful to compare it to frame relay. Frame relay relies on a CIR to indicate the guaranteed bandwidth of each PVC. Instead of a CIR, ATM relies on a *sustainable cell rate* (SCR), which defines the guaranteed cell rate that a station can transmit into the network. On the other hand, the *peak cell rate* (PCR) is analogous to frame relay's EIR, defining the maximum traffic rate acceptable by the network. Finally, the *maximum burst size* defines the maximum cell volume acceptable within a defined period of time, usually one second.

Based on the characteristics described previously, ATM defines a series of service classes that accommodate the requirements of all application types. *Constant bit rate service* (CBR) defines a CoS that emulates the constant-delay, high-priority nature of a dedicated, private-line circuit. *Variable bit rate service* (VBR) guarantees that traffic falling within the SCR will be carried, while anything above the SCR will be marked DE and may be discarded if it exceeds the maximum burst size.

Variable bit rate—real time service (VBR-rt) is similar to VBR, but adds better congestion control and latency guarantees. It is typically used for the transport of voice and video, which require minimal delay through the network.

Available bit rate (ABR) is a variation on a theme; although it does not make any guarantees relative to delay, it does guarantee bandwidth availability. In ABR, cells may be buffered by the network, and in some cases, the network may attempt to flow-control an offending device. As a result, ABR is something of a hybrid of CBR's guaranteed transport and VBR's flexibility, with a bit of UBR (described in the following) thrown in for good measure.

Finally, *unspecified bit rate service* (UBR) makes no guarantees at all, similar to IP. In spite of what may appear to be a less-than-desirable service choice, UBR is widely deployed.

CONGESTION CONTROL IN ATM

Flow management in ATM has been widely studied, and two principal mechanisms, based on standards, have resulted. The first is the use of the *explicit forward congestion indication* (EFCI) bit found in the ATM cell header, which is used to notify downstream switches of the presence of congestion. When a switch receives a cell that has the EFCI bit turned on, the receiving switch responds by sending a choke cell to the transmitter (known as a *resource management cell*) that attempts to slow the transmission volume. The alternative flow control technique is called *explicit rate,* a more elegant scheme than the rather primitive EFCI solution. When explicit rate is implemented, a *resource management cell* is inserted in the cell stream after every thirty-second cell. These indicator cells enable receiving ATM network components to respond more quickly because they can detect a pattern in the information contained in the RM cells that might indicate a buildup of congestion.

Attentive readers may have noticed that although ATM has an EFCI, there is no EBCI. The reason for this is that congestion control is done on a *virtual channel* (VC) basis, and because VCs are unidirectional, there is no need for backward congestion notification.

SO: WHY ATM?

If we were to design the perfect network, it would

- Operate at a range of speeds
- Exhibit a low error rate over long operational periods
- Handle multiservice and multiprotocol support
- Be based on international standards
- Support bandwidth-on-demand
- Be imminently scalable

- Manage traffic well during periods of congestion or other disruptive events
- Be interoperable with other technologies
- Offer diverse QoS
- Be future-proof
- Have a well-developed NM scheme

Let's consider ATM, given the list of capabilities shown previously.

Operate at a range of speeds. ATM is part of the broadband ISDN protocol model, designed to operate at SONET/SDH speeds. This includes bandwidth provisioned from OC-1 (51.84 Mbps) to OC-192 (9.95 Gbps). In the unlikely event that more bandwidth is needed, DWDM can provide multiples of SONET speeds.

Exhibit a low error rate over long operational periods. ATM is a fast packet technology, which by definition experiences extremely low incidences of error. Furthermore, it interfaces directly with SONET, which as an optical technology is equally free of the typical errors that plague metallic facilities.

Handle multiservice and multiprotocol support. ATM's well-defined service classes (CBR, VBR, ABR, and UBR), as well as the diverse service and parameter qualifiers that the protocol stack relies on (timing sensitivity, bit rate variability, connection mode, sensitivity to burstiness, and so on), give it the unique capability to not only transport a wide variety of service types, but to guarantee that the delivery quality customers expect based on the nature of the application they derive from is assured.

Be based in international standards. Both the ITU-T and the ATM Forum have contributed to the development of standards for ATM that are recognized worldwide.

Support bandwidth-on-demand. The well-developed signaling scheme that underlies the ATM protocol model enables the allocation of bandwidth on an as-needed basis. Because SONET is often the assumed physical layer, there is never a question about the availability of bandwidth when an application requires it.

Be imminently scalable. Because ATM appears to be assuming the role of a primary core technology for the next-generation network, manufacturers of ATM switching equipment have made scalability a key tenet of their design efforts. Like central office switches, ATM switches can be augmented to respond to customer growth as required.

Manage traffic well during periods of congestion or other disruptive events. Because ATM is much more than an asynchronous data transport technology, it must be able to handle the mixed performance requirements of both delay-sensitive and insensitive services. Consequently, ATM supports a variety of NM protocols that have the capability to perform flow control and to route around problem areas in the network should the need arise.

Be interoperable with other technologies. ATM is not particularly discriminating as technologies go and is in fact rather forgiving. It has the capability to emulate other technologies and can thus provide transport between them when required. For example, ATM offers a popular service called *frame relay bearer service* (FRBS). With FRBS, service providers can deploy an ATM network from which they can provision both ATM and frame relay services. This is accomplished with ease. If a customer wants to buy frame relay service from the provider, the provider simply adds frame relay interface cards to the ATM switch (see Figure 1-39). The customer sends frames to the switch, which converts them to cells for long-haul transport. At the other end of the circuit, the ATM switch converts the cells back to frames for delivery to the intended recipient. In this situation, the customer has no idea that their traffic is actually being transported by a far more efficient network fabric than ATM; the good news is that they don't *have* to know. From a services point of view, it doesn't matter.

Offer diverse QoS. The primary reason that ATM has emerged as the recommended foundation technology for the deployment of IP networks is its capability to offer service

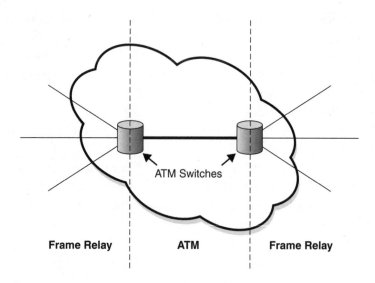

FIGURE 1-39   Emulation in ATM.

granularity. No one will argue with the observation that ATM is *not* an ideal transport choice for voice, or video, or medical images, or data. Clearly, circuit switching offers a better solution for voice, while a private-line circuit is much more appropriate for the rigorous demands of video or medical imaging files. ATM, however, provides the ideal solution for the broad mix of all of those services. However, if a service provider wanted to deploy a single-fabric network that could satisfy the bulk of its customers' needs while taking into account cost-effectiveness, ATM would be the only contender.

Be future-proof. ATM is. There is no technology looming on the immediate horizon that threatens ATM's supremacy. Many believe that IP will ultimately be retrofitted with the capability to provide diverse QoS, but until that happens, ATM is the only answer that makes complete sense.

Have a well-developed NM scheme. ATM NM offers well-designed maintenance, monitoring, and provisioning capabilities. When combined with the inherent capabilities of SONET, the result is a well-run network.

THE ATM MARKET

For a long time, ATM's position in the marketplace was sketchy because it was considered an expensive alternative, especially in the local area. Over time, however, its proper position has become clear. ATM provides dependable, high-volume bandwidth for large-scale enterprise networks. In smaller corporate networks, it provides an aggregation solution for lower-bandwidth technologies such as T-1 and DS-3. Today, the cost for ATM service has dropped to the point that it is approaching parity with frame relay. Most interexchange carriers offer ATM service at prices comparable to frame relay and in some cases cheaper. Interexchange carriers by and large charge slightly more for ATM, but expect to offer the two at similar prices very soon.

Both PVC and SVC services are available from ATM, but PVC service is far more widely deployed. As in frame relay, SVCs are difficult to measure, track, and bill and are therefore eschewed by most carriers. Some do offer the service, however; Sprint and Qwest bill their SVC customers according to cell volume, while AT&T bills are based on connection duration.

ALTERNATIVES TO ATM

We have already mentioned the fact that ATM may someday be ousted by a more QoS-capable IP protocol. That is not likely to happen in the near-term, but two alternatives have emerged that could prove to be viable challengers.

Cisco has released its *spatial reuse protocol* (SRP) as well as devices that implement it. With SRP, routers can be interconnected by dual fiber rings using SONET and/or DWDM. The rings offer redundancy and operate at 10 Gbps with a switchover speed of less than 50 ms. In metropolitan area network scenarios, SRP could dramatically lower the cost of network deployment for corporations or ISPs looking to interconnect multiple sites by eliminating the need for an expensive ATM infrastructure. The one thing SRP does not adequately offer today, however, is a well-defined strategy for QoS, something ATM does extremely well.

Nevertheless, SRP has found a home in the optical world because of its use in a new offering called *resilient packet ring* (RPR). RPR is a network transport scheme for fiber rings. The IEEE began the 802.17 RPR standards project in December 2000 with the intention of creating a new Media Access Control layer. Fiber rings are widely used in both metro and wide area networks; these topologies, unfortunately, are dependent on protocols that are not scalable for the demands of packet-switched networks such as IP. The 802.17 RPR working group promotes RPR as a technology for device connectivity. Issues currently being addressed are bandwidth allocation and throughput, deployment speed, and equipment and operational costs.

A related technology is *optical burst switching* (OBS). OBS supports real-time multimedia traffic across the Internet, a natural pairing because the Internet itself is bursty. OBS is particularly effective for high-demand applications that require high-bit-rate, low latency, and short duration transmission characteristics.

OBS uses a one-way reservation technique so that a burst of user data, such as a cluster of IP packets, can be sent without having to establish a dedicated path prior to transmission. A control packet is sent first to reserve the wavelength, followed by the traffic burst. As a result, OBS avoids the protracted end-to-end setup delay and also improves the utilization of optical channels for VBR. By combining the advantages of optical circuit switching and those of optical packet or cell switching, OBS facilitates the establishment of a flexible, efficient, and bandwidth-rich infrastructure.

Another ATM alternative is *Dynamic Synchronous Transfer Mode* (DTM). Introduced by three Swedish firms in January 1999, DTM is similar to circuit switching in that it relies on timeslots, but is different in that it can reallocate those timeslots on an as-needed basis to meet the demands of bursty applications when they arise.

ATM SUMMARY

ATM has established itself as a reliable and flexible core switching technology for the vast majority of service types. It relies on international standards, provides good QoS and CoS control, has a robust embedded NM scheme, and is future-proof enough to handle the growing demands of new applications for some time to come. What it lacks, however, is a widely accepted and flexible addressing scheme and the universality that other protocols enjoy. It is an expensive solution when deployed at the edge of the network, but is quite cost effective when used as a core, WAN technology. Ideally, then, ATM, combined with another protocol that overcomes its few shortcomings, would be an unbeatable combination.

IP

IP is part of the well-known TCP/IP protocol suite that came about in the 1960s as an integral part of the Department of Defense's ARPANET project. It is a network layer protocol, responsible for routing functions and congestion control, and works closely with its transport layer counterpart, the *Transmission Control Protocol* (TCP). TCP/IP supports a wide array of application services and will operate over many different physical media types.

IP is a connectionless network layer protocol, which means that it does not perform a call setup prior to transmitting data packets. Instead, it transmits the packets into the network, each bearing a complete destination address, and trusts the network to deliver them to their final destination. With the help of various routing protocols the packets do generally arrive, but because the service is connectionless and because connectionless networks do not discern a relationship between packets that originate from the same source, they may take different routes from the source to destination based on changing network congestion and other factors. Consequently, they may arrive out of sequence. The advantage of a connectionless network, of course, is that it has the capability to avoid troubled areas in the network by dynamically routing around them. This flexibility is worth the cost of connectionless service, particularly because the transport layer (TCP) will generally correct any discrepancies that arise from IP's connectionless nature.

ADDRESSING IN IP

The current version of IP, known as IP Versions 4 (IPv4), relies on what is known as a *dotted decimal addressing scheme*. The name derives from the fact that IP addresses comprise four 8-bit segments separated by periods, as shown in the following:

255.255.255.255

Each IP address is 32 bits long (four 8-bit components), which means that 2^{32} (roughly 4.2 billion) possible addresses can be created. The good news is that 4.2 billion is *a lot*. The bad news is that it isn't enough! Because of the immense and unexpected popularity of IP as a universal addressing scheme, we are getting dangerously close to exhausting the available address space. There are a number of reasons for this: The diversity of devices that can actually be addressed by IP is growing, networked devices in general are growing, and the manner in which IP addresses are assigned is not particularly efficient.

To understand the challenges that face IP addressing, it is first necessary to understand how the addresses work. There are five classes of IP addresses, labeled A, B, C, D, and E. Classes A through C are used commercially, while Class D is reserved for multicasting and E for future use. They are created as follows. In any given company, it is necessary to identify both the network that serves that company as well as the devices attached to that network. By assigning the four 8-bit pieces of the IP address in a variety of ways, we can create a tiered address scheme that enables us to identify a small number of addresses with lots of hosts (devices), a large number of networks with a few hosts, or anything in between. Consider the illustrations in Figure 1-40. In a Class A address, three of the bytes are assigned as host identifiers, while one is used to identify networks. Consequently, a Class A address can identify as many as 16 million unique hosts (2^{24}) and up to 126 ($2^7 - 2$) unique networks. As it turns out, not all of the bits in each byte are actually used for addressing—some are reserved to indicate the Class of the address. For example, in Class A, the most significant bit is set to 0 to indicate that it is, in fact, a Class A address, leaving seven bits to identify the network. Some addresses are reserved, so the maximum number of identifiable networks in a Class A address is 26. Similarly, a Class B address can identify 65,534 hosts on 16,382 networks, while a Class C address, the most commonly deployed of all, can identify 254 hosts on each of more than 2 million networks.

The biggest problem facing IP today is the fact that when IP addresses are assigned to corporations, they are generally assigned an entire address block based on the number of employees/devices they must provide with a static network address. For example, a company of 50 people might acquire an entire Class C address, which would allow them to easily accommodate their 50 employees but would leave a significant number of IP addresses within the range idle. This is a serious problem; some reports estimate that as few as 20 percent of all assigned IP addresses are actually in use.

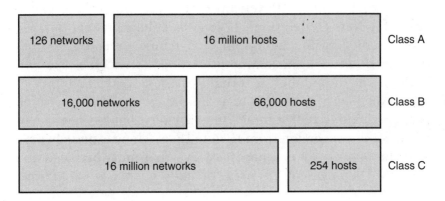

FIGURE 1-40   IP addressing scheme.

To combat this problem, a number of solutions have been devised. The first we will describe is called *subnet masking*. In subnet masking, IP addresses are broken into three pieces instead of two (NETID and HOSTID). In subnet masking the first piece identifies the network, the second, a subnet address, and the third, the host on that subnet. Each subnet is limited to 254 nodes, but within a Class B address, for example, there can be as many as 256 uniquely identifiable subnets. In effect, the network is subdivided into pieces that make management and administration easier—and far more granular. By creating a larger number of unique networks within the address space normally reserved for one network, a better assignment of IP addresses is possible.

The second technique that is being used is a protocol called *Dynamic Host Configuration Protocol* (DHCP). DHCP enables a server to dynamically assign IP addresses to requesting stations, thus eliminating the problem of unused IP address space.

The third solution is called *Classless InterDomain Routing* (CIDR). CIDR (pronounced like the drink) enables Class C addresses to be broken into smaller pieces than 254 hosts. This helps to resolve the problem of IP address overassignment. If a company buys a Class C block, for example, and later determines that it needs another 50 addresses, it must actually acquire another entire Class C address. With CIDR, they can buy small pieces.

Finally, the solution that many are waiting for is the next version of IP, called IP Version 6 (IPv6).[22] Sometimes called *IPng* (for Next Generation), IPv6 adds enhancements to IPv4, including 128-bit addressing, dynamic reconfiguration, enhanced network security, multicasting, better QoS discrimination, and a number of other features. With 128 bits of addressing space, 2^{128} unique addresses can be created, which is approximately 10^{38} in total. One difficulty that has been identified is the conversion from IPv4 to IPv6. Many experts believe that the conversion is a necessary evil that must be faced and that the deployment of *network address translation devices*

---

[22]IPv6 is described in IETF RFC 1752.

(NATs), which will convert between IPv4 and IPv6 addresses, will be required for some time. Indications from the so-called early adopters and implementers of IPv6 are that the advantages far outweigh the downsides of the conversion.

WHY IP?

Of all the questions answered in this book, "Why IP?" is most pressing today. IP has enjoyed more press and attention than any other technology topic in the last few years and for good reason.

One of the biggest problems that network services providers have traditionally faced is the issue of *divergence*. They offer a wide range of services including POTS, ISDN, X.25 packet transport, frame relay, ATM, high-speed point-to-point services based on SONET, and many others. They offer them because customer applications demand them.

The problem with having such a diverse product set is that it requires an equally diverse support infrastructure. Each of these technology-based services requires billing, provisioning, installation, administration, maintenance, troubleshooting, repair, and direct customer contact support. This divergence is expensive, cumbersome, time-consuming, and *not* in the best interest of either the customer or the service provider.

Imagine a scenario, then, where a service provider could deploy a single network fabric from which they could provide all of their offered services, guarantee high quality, enjoy lower costs in every facet of the business, and easily meet the evolving needs of all customers. In 1982, that is precisely the role that the service provider (all one of them) played. If a customer came to AT&T with a business problem in search of a telecommunications solution, AT&T would quite comfortably, and reliably, respond, "Look: You just tell us where the terminals are, and where the host is, and where the bank branches are, and we'll take care of the stuff in the middle. That's our job." As a service provider, they did not burden the customer with cloud-related stuff because they *knew* that their network could reliably meet the demands of the customer, regardless of the nature of those demands.[23] The customer had no reason to go anywhere else because they knew that AT&T would deliver a solution that worked.

The fact of the matter is that service providers need to get back to that 1982 model. OK, not the monopoly model, but a model that places them in the position of being a service provider that is *so good at providing service that the customer wouldn't* think *of going anywhere else.* One way to do that is to build a communications delivery infrastructure that converts *divergence* to *convergence.* If service providers can deploy a technology base that dramatically simplifies the manner in which they deliver services, then untold advantages result, including those listed previously—lowered costs, shorter time to market, and diverse services.

MAKING IT HAPPEN

Let's go back and revisit our perfect network model once again. Earlier, we said that the ideal network would

- Operate at a range of speeds
- Exhibit a low error rate over long operational periods
- Handle multiservice and multiprotocol support
- Be based on international standards
- Support bandwidth-on-demand
- Be imminently scalable
- Manage traffic well during periods of congestion or other disruptive events
- Be interoperable with other technologies
- Offer diverse QoS

---

[23]I am not looking to start a debate here on the merits or evils of monopoly business; just consider the service implications of what I am describing.

- Be future-proof
- Have a well-developed NM scheme

As we demonstrated, ATM as a lower-level switching fabric uniquely satisfies these requirements, but there are at least three others that ATM does not provide that must be considered. Both derive from a higher-layer network protocol.

First, the network protocol employed must be widely understood, universally available, inexpensive to implement, and simple to manage and administer. Second, it must offer an addressing scheme that will support large and growing population bases and address a wide variety of device types. Third, it must be universally application-friendly.

IP meets these requirements quite effectively. It is one of the most widely deployed protocols on Earth, found in LANs, MANs, and WANs, as well as in every modern network operating system. In terms of ubiquity, IP has no equal. Furthermore, because of its universal acceptance, it is widely understood, and sophisticated tools exist to manage networks based on it.

As an addressing scheme, IPv4 certainly has its weaknesses. However, they are rapidly disappearing under the onslaught of such solutions as DHCP, CIDR, subnet masking, and, of course, the inevitable arrival of IPv6. The result of all this is that IP has become the only universal addressing scheme we have. It is accepted worldwide, has more than enough capacity, and, thanks to efforts by a plethora of vendors, is now interoperable with network operating systems, signaling systems, and most applications that require it.

IP/ATM INTEGRATION

So what's the conclusion we're looking for? If a service provider wants to deploy a full-service network, a likely solution will comprise an ATM core with IP running on top of that core. By combining the advantages of ATM with those of IP, service providers can create a universal cloud that really does do it all, from end-to-end. IP brings the statistical loading advantages of a packet network as well as addressing and universality; ATM adds QoS, bandwidth, and standards-based technology in the transport core. With this combination of technologies, network providers can make available every possible service including dedicated private line, frame relay, ATM, voice, asynchronous data transport, interactive and distributive video, and many others. The elegance of this solution is that the underlying technology need not concern the customer, because ideally, the service provider no longer sells technology—they sell service, which relies, of course, on technology. The customer, however, is not burdened with the details.

Consider the following scenario: Some day in the not too distant future, XYZ ILEC decides that they are technologically ready to convert to an all-IP network. They have relied on ATM for a long time to provide the solid QoS support needed to satisfy customer demands, but now new QoS protocols (discussed in the next chapter) have advanced to the point that ATM is no longer necessary in every case. Those QoS protocols, combined with IP and a high-speed, robust physical infrastructure, obviate the need for ATM's considerable capabilities. So, late one night, at a preappointed time, XYZ's central office technicians take a deep breath and pull the switch, converting the company to an all-IP infrastructure. And the best part of all this? *The customer never knows it has happened.* Services go in; services come out. The nature of the underlying fabric is completely invisible to the customer—as it should be. Service quality does not change from a customer's point of view, but from the service provider's perspective, life just became remarkably more simple. A single network now does it all.

The evolution to bring this about is already underway, and major vendors are lining up to meet the demand and become the center of the universal services cloud. Lucent Technologies, for example, offers the 7 R/E, a potential successor to the 5ESS™. The design of the 7 R/E recognizes the enormous embedded value of the 5E platform, but also recognizes that circuit switching is no longer the answer to every problem. Built around an IP/ATM core, the switch combines the stability and latency-free service of circuit switching with the robustness, flexibility, and service-transparency of IP. ATM serves as the core transport mechanism, but as our short story in one part described, the time may come when ATM is no longer required, at which point IP may take over. The 7 R/E is fully capable of enabling that to happen when the time comes.

As the name implies,[24] the 7 R/E can be used as an immediate full-service switching solution for green field networks[25] or as an evolutionary solution for legacy installations. In general, the 7 R/E enables the embedded 5E base to continue with its evolution because in spite of claims to the contrary, circuit switching is still a viable, growing solution for voice applications. As the network continues to advance, the 5E can be evolved to a 7 R/E as required, but the design of the switch enables the evolution to be graceful, modular, and timely. This enables service providers to maximize the utility of their investment in circuit switching while at the same time moving forward with packet-based IP. Finally, an integral NM system completes the package.

Nortel Networks has a similar plan underway with its Succession™ product line. The platform facilitates the convergence of voice and data, transforming the traditional circuit-switched network into a distributed broadband network that delivers seamless services over an IP/ATM backbone optimized for the transport of voice traffic, thus ensuring the protection of highly lucrative legacy services while at the same time creating a viable platform for the delivery of broadband. The switch is based on international standards and offers many advantages, including compatibility with multiple vendors, reduced cost of operation and ownership, rapid, smooth evolution to packet services as required, and increased trunk capacity. Additionally, the switch is largely future-proof, fully capable of handling evolving applications.

These manufacturers understand the direction that their clients are going and also understand the demands being placed

---

[24]The R stands for Revolution, the E for Evolution.

[25]A green field market is one that is new and does not yet have an embedded infrastructure.

on them by *their* clients. They are building products that satisfy the requirements of the third tier, their customers' customers. They are also helping service providers create the network that will position them to make the leap to being true *service providers*.

SUMMARY

In this section we dissected the network transport cloud and its access extremities. We started with a discussion of access technologies, including 56K modems, ISDN, cable modems, wireless solutions, and DSL. Once we selected a method for accessing the cloud, we examined the technologies that make up the cloud itself: circuit, packet, and fast packet switching, including SONET, DWDM, frame relay, and ATM.

Customers rely on these technologies in various combinations to resolve business problems. They are chosen based on bandwidth requirements, cost, the capability to provide required levels of service quality, local availability, and the degree to which they will interoperate with other technologies and applications.

By and large, customers are not in the technology business and don't want to be. They have their own business issues to deal with and do not care about selecting the best technology mix to manage those issues. What they want is for the service provider to do it: to understand their business challenges, to offer a solution, and to not only have it work, but to know without a second thought that it is the *best* solution.

The key to providing that best solution lies in the ability to provide services. It is easy to believe that services come in a limited number of flavors with names like voice, video, image, and data transport. If a service provider believes that their business is to simply provide access and transport, then they will be remanded into the corner with the other unimaginative commodity providers. The fact is that customers can buy their access and transport services from a broad spectrum of suppliers today, and recent legal rulings not only make that possible, they make it *easy*. Commodity providers win the game if they have the cheapest price for their commodity among all the other commodity providers. Perhaps I'm naïve, but that doesn't sound like a service to me. Dictionaries define service as "an act of assistance or benefit to others." Service providers, then, should be providing benefit, and in the eyes of customers the best benefit they could possibly deliver is some form of competitive advantage. Please understand: Access and transport are critical components of modern business today. Without the network, *there is no business.* However, imagine the power of a full-service network that not only provides access and transport but also has the capability to adapt to customer business requirements. Now *that's* a service.

In the next section, we will examine the convergence of companies, the second component of the convergence troika. As we observed in the introduction, there is something of a feeding frenzy underway as companies in the industry work to reinvent themselves to become the *first best provider* of services for their customers. They realize that they do not have everything it takes to do it all and also understand that they do not have the time nor the wherewithal to develop additional capabilities in-house. Were they to try, the market would leave them behind in a cloud of dust. Instead, they combine their efforts with those of other providers, thus rounding out their collective abilities. In this next section, we will describe that phenomenon.

Finally, in the last section of the book, we will discuss the basic services that customers want, including voice, video, image, data, fax, and e-mail, but we will not stop there. Service convergence means logically integrating those basic capabilities to create new and exciting applications that use the underlying technology to innovatively resolve business challenges. That integration process requires other technological components, including signaling systems, QoS protocols, and NM. The resulting applications are powerful and compelling and have significant impacts on the businesses that implement them. They include IP voice, unified messaging, virtual private networks, and voice-enabled Web sites.

Remember: Knowledge about a customer's business processes leads to the development of solutions through technology convergence. Technology convergence in turn leads to company convergence as competitors vie to have the most complete solution in as short a timeframe as possible. Services convergence, the result of the first two, represents the fruits of the labor of technology and company convergence, and that's the end game.

PART TWO

COMPANY CONVERGENCE

I was to learn later in life that we tend to meet any new situation by reorganizing, and a wonderful method it can be for creating the illusion of progress while producing confusion, inefficiency, and demoralization.

—Petronius Arbiter, Emperor Nero's first-century advisor on issues of luxury and extravagance

In Part One, we discussed how knowledge about the customer helps network providers select the technologies that will help them satisfy the customer's requests for service. We also discussed the fact that technology convergence is only a piece of a rather complex puzzle. The second piece, company convergence, must occur in lockstep if service convergence is to take place.

When service providers assess their roles in the modern telecommunications marketplace, they often discover that the game has changed so radically that they don't know the rules anymore. Customer expectations have shifted, lines of business have become blurred, players have changed, and the roles that were so well defined for each of them seem foreign. Today, with the industry undergoing what many have described as a technological meltdown, players in the telecommunications game have become particularly aware of the need to change the way they interact with customers and live out their specific roles.

The tradition-bound model of manufacturer selling to wholesaler, wholesaler to retailer, and retailer to consumer is rapidly disappearing as e-commerce and e-business change the model of doing business, and with it the competitive landscape. The stately, linear business processes that have characterized corporate practices for a long time are giving way to faster, sleeker models, driven by a demanding customer base that is not afraid to exercise its right of competitive choice. The Internet, perhaps the greatest agent of change in business history, is redefining the market and forcing a redefinition of *the business of doing business*. The phenomenon of doing business electronically was enhanced in the late 1990s as the Internet was discovered and the world became comfortable with it; later it became what many have described as an indispensable tool for shopping, browsing, doing research, and conducting *business-to-business* (B2B) and business-to-consumer commerce.

According to a study by the Gartner Group, B2B spending will total more than $8.5 trillion by 2005, a significant number. For the last five years, B2B has enjoyed a growth rate in excess of 80 percent per year, and although the current economic slowdown has resulted in a reduction of overall commerce, including B2B, the growth continues apace and will continue to grow as the economics shift and as consumers and enterprises alike continue to discover the advantages that can be had from electronic transactions.

As in Part One, a summary will now be presented of the key areas of activity within the not inconsiderable domain of company convergence. As the players morph and recreate themselves in response to the dynamics of the industry and evolving customer demand, the landscape changes. There are numerous indicators of the continuing growth of the telecommunications marketplace that make plenty of room for optimism among investors. These include the following:

- The numbers of worldwide Internet subscribers are expected to increase from 404 million in 2001 to 480 million in 2002, and to 550 million by 2003.

- Worldwide broadband subscribers, including cable modem, *digital subscriber line* (DSL), and wireless, will increase from 22 million in 2001 to 35 million in 2002, and to 52 million in 2003.

- Finally, U.S. backbone traffic is expected to increase from 2.4 exabytes in 2001 to 3.7 exabytes in 2002, and to 5.9 exabytes in 2003. How much is an exabyte? A lot.

So, what are the key indicators that will guide the further convergence of companies within the technosphere? Read on.

THE PLAYERS IN THE GAME WILL CONTINUE TO EVOLVE

The recreation of the telecommunications industry that began with the divestiture of AT&T in 1984 reaches far beyond the borders of the United States. It has become a global phenomenon as companies move to compete in what has become an enormously lucrative business. Customers, however, have been quick to point out that technology is a wonderful thing, but unless the provider positions technology in a way that makes it useful to the would-be customer, it has zero value.

Customers expect account representatives to have strong technical knowledge as well as knowledge about them as customers: business issues, competition, major segment concerns, and so on. Furthermore, the level of customer satisfaction or dissatisfaction varies from industry segment to industry segment. *Incumbent local exchange carriers* (ILECs) are too rigid and expensive. *Competitive local exchange carriers* (CLECs) must compete on price, but must also show a demonstrated ability to respond quickly and offer *operation support systems/ network module* (OSS/NM) capabilities. Ninety-two percent of all business customers want to work with an account team when purchasing services. However, account teams by-and-large are not creating long-lasting relationships or loyalty. Eighty-eight percent of those same customers say that they buy their services from the local provider, but only 53 percent say that they would continue to use them if given the choice. According to numerous studies, the single most important factor that customers take into consideration when selecting a service provider is cost. Yet the most important factor identified for improvement is customer service.

A NEW REGULATORY ENVIRONMENT IS NEEDED AND WILL BE CREATED

The greatest challenge that has historically faced service providers is the requirement to provide true universal service, both in rural areas as well as in metropolitan and suburban areas. The American Communications Act of 1934, signed into law by President Roosevelt, mandated that telephony service would be universal and affordable. In order to make this happen, AT&T agreed in the 1940s to offer low-cost service with the tacit approval of the Justice Department to charge a slightly higher price for long distance, some business-line services, and value-added services as a way to offset the cost of deploying high-cost rural service and to support low-income families. These subsidies made it possible to offer true universal, low-cost service.

Today 60 to 70 percent of the local loops in the United States are still subsidized to the tune of anywhere from $3 to $15 per month. As a result, the incumbent service providers often charge significantly less for the service they sell than it actually costs them to provide it.

The problem with this model is that only the ILECs have the right to enjoy these subsidies, which means that CLECs are heavily penalized right out of the starting gate. As a result, they typically ignore the residence market in favor of the far more lucrative business markets. Business markets are significantly easier and less costly to provision than residence installations because of the dominance of *multitenant units* (MTUs).

The current regulatory environment in many countries does not address the disparity that exists between incumbent providers and would-be competitors. Recent decisions, such as the United States' Communications Act of 1996, have consisted of a series of decisions designed to foster competition at the local-loop level. Unfortunately, by most players' estimations, it has had the opposite effect.

New technologies always convey a temporary advantage to the first mover. Innovation involves significant risk; the temporary first-mover advantage allows them to extract profits as a reward for taking the initial risk, thus allowing innovators to recover the cost of innovation. As technology advances, the first-mover position often goes away, so the advantage is fleeting. The relationship, however, fosters a zeal for ongoing entrepreneurial development.

Under the regulatory rule set that exists in many countries today, incumbent providers are required to open their networks through element unbundling and to sell their resources—including new technology—at a wholesale price to their competitors. In the minds of regulators, this creates a competitive marketplace, but it is artificial. In fact, the opposite happens: The incumbents lose their incentive to invest in new technology, and innovation progresses at *telco time*. Under the wholesale unbundling requirements, the rewards for innovative behavior are socialized, while the risks undertaken by the incumbents are privatized. Why should they invest and take substantial economic risks when they are required by law to immediately share the rewards with their competitors?

Ultimately, the truth comes down to this: Success in the access marketplace, translated as sustainable profits, relies on network ownership—period. Many would-be competitors such as E.spire and ICG bought switches and other network infrastructure components, built partial networks, and created business plans that consciously relied on the ILECs for the remainder of their network infrastructure. Because of the aforementioned subsidies, they quickly learned that this model could not succeed, and many of them have disappeared.

Consider this: When WorldCom bought *Metropolitan Fiber Systems* (MFS), they weren't after the business; they were after the in-place local network that MFS had built and that would enable WorldCom to satisfy customer demands for end-to-end service without going through the cost of a network build-out. They paid more than six times the value of MFS' in-place assets for the company. The message? Network ownership is key.

Two major bills currently before the U.S. Congress (Tauzin-Dingell, HR 1542, and Cannon-Conyers, HR 1697) aim to reposition the players. 1542 makes it easier for ILECs to enter the long-distance, Internet, and data businesses. 1697 tightens controls over ILECs' entry and applies significant fines for failure to open incumbent networks to competitors. 1542 has already passed one subcommittee, but a major battle is expected between the supporters of the two bills. Most industry analysts believe that the industry needs less government controls and more competition; the FCC appears to agree.

SOLUTIONS

So, what are the possible solutions? One is to eliminate the local service subsidies and allow the ILECs to raise rates so that they are slightly above the actual cost of provisioning service. Subsidies could continue for high-cost rural areas and low-income families (roughly 18 percent of the total), but would be eliminated elsewhere. New entrants would then have a greater chance of playing successfully in the local access business. The subsidy dollars, estimated to be in the neighborhood of $15 billion, could then be redeployed to finance universal broadband access deployment. The monies could be distributed among ILECs, CLECs, cable companies, *Internet service providers* (ISPs), wireless companies, and DSL providers in order to facilitate broadband.

A second solution is to call for the structural separation of the ILECs, which would result in the establishment of retail and wholesale arms. The retail arm would continue to sell to traditional customers, while the wholesale arm would sell unbundled network resources to all comers. The result would be enhanced innovation; the downside would undoubtedly be strong resistance from the ILECs.

Many believe that high-speed data and Internet access should have been exempted from the Communications Act of 1996's mandates, and that it should have targeted only traditional voice services where the ILECs have clear monopoly positions. They contend that unbundling and wholesale requirements should not apply to nonvoice services.

The new regime at the FCC, led by Michael Powell, has indicated that it wants less government intervention in the telecommunications marketplace rather than more, a good thing in light of the current industry. One could argue that well-intended regulatory strictures have in fact done damage.

Consider the case of WorldCom. The company's original plan in the late 1990s was to challenge local service providers all over the world by creating a broadband voice and data *Internet protocol* (IP) network through acquisitions and mergers. The regulators, concerned by WorldCom's aggressive plans, felt that the intended company looked too much like a monopoly. They forced the divestiture of MCI's Internet company to Cable & Wireless and rejected the proposed merger with Sprint because of fears that they would control 80 percent of the long-distance market. This decision was made while long-distance revenues were plummeting due to the influence of such disruptive technologies as multichannel optical transport and IP. The result is two badly weakened companies that have not yet recovered—and may not. In fact, they could be prime acquisition targets for the ILECs. At the time of this writing, AT&T is in talks with BellSouth over a proposal to sell AT&T's wireline long-distance assets to the company.

Another example is AT&T itself. There was a huge expectation that AT&T would be a big winner in the local broadband access game following its acquisition of cable properties for its plans to deliver high-speed Internet and interactive services. Many analysts expected a market cross-invasion between the ILECs and cable providers, but it never happened. Cable providers concentrated on adding Internet service and additional channels, and telephone companies concentrated on penetrating the long-distance market. Furthermore, when talk of open access and loop unbundling began to be targeted at the cable industry in 2000, AT&T's hopes of a competitive advantage through cable ownership were dashed.

There were also expectations that ILECs would work hard to penetrate each other's markets, but this never happened. Who better than the ILECs knows that network ownership is the most critical factor for success in the local access game? If you control the network, you control the customers. More importantly, if you *don't* have that control, *don't get into the game.*

E-COMMERCE AND E-BUSINESS WILL STRENGTHEN AND MATURE

E-commerce and its descendant, e-business, have emerged as the principal drivers behind the continued fame and success of the Internet. They have revolutionized purchasing, *customer relationship management* (CRM), and supply chain management. They have also provided the functional basis for the data-mining and knowledge management activities that characterize today's successful company.

E-COMMERCE

There is nothing remarkable about the process of making an online purchase. In fact, it is not all that different from making a purchase by telephone. Other than the fact that the telephone agent is replaced with a web server that performs the same functions, the process is identical. Yet e-commerce has become one of the most influential phenomena of the Internet era.

So, what is it exactly? Opinions differ greatly on a precise definition. During a Senate committee hearing on pornography associated with the Communications Act of 1996, one of the ranking senators on the committee was asked to state his own definition of pornographic material. After careful thought and a certain amount of fluster, he responded, "Look, I can't define it, but I know it when I see it." E-commerce is equally hard to define, but when implemented, its impact is as hard as stone. To many, e-commerce consists of the process of ordering merchandise by phone from a mail-order company, sending a fax to the diner next door to order lunch, or ordering a pizza from Pizza Hut. To others, it is the use of *electric data interchange* (EDI) transactions between automobile manufacturers and their vendors to carry on the business of manufacturing cars, or a process by which a service provider sends an electronic bill to a large customer in lieu of a paper invoice.

All of these are versions of e-commerce, but today e-commerce is usually taken to mean the process of shopping online, selecting a product, securely paying for the product, and ensuring that the customer receives his or her purchase within a reasonable amount of time. In all cases, it involves the use of secure payment protocols to protect the buyer, and in some cases it may involve the use of electronic money to protect his or her anonymity.

WHY E-COMMERCE?

There are numerous reasons for the success of e-commerce, including convenience, lower cost, access to global product markets, and anonymity. There are also compelling economic reasons for its success. A June 1999 study funded by Cisco and conducted by the University of Texas demonstrated that online commerce is not a passing fad. In 1998 alone, the Internet economy generated $301 billion in U.S. revenues and was directly responsible for 1.2 million jobs. It also indicated that Internet workers are, on average, 65 percent more productive than non-Internet workers. Furthermore, the average revenue per Internet employee was approximately $250,000 compared to $160,000 for the non-Internet employee. At $300 billion, the Internet economy rests in the same lofty heights as the automotive industry ($350 billion) and the telecommunications market ($270 billion). Today those numbers have grown significantly and continue to grow at a rapid pace. In a separate study, Forrester Research observed that in 1998, B2B e-commerce was a $43 billion dollar industry, but will climb to $1.3 trillion by 2003.

According to a study by Dr. Lawrence Roberts, considered one of the founding fathers of the global Internet, U.S. Internet traffic on core IP service providers' networks increased fourfold between April 2000 and April 2001. Internet traffic is doubling every six months on average, which compares to an average growth rate of 2.8 times per year since 1997. The study, which sampled network traffic from the top 19 U.S. data carriers in April 2000, October 2000, and April 2001, caused Roberts to conclude that 80 percent of U.S. Internet traffic is generated by businesses, and that international traffic is still growing at the "pre-2000 growth rate of 2.8 times per year." The research also concluded that 50 percent of all Internet traffic is carried by the top four ISPs.

FROM E-COMMERCE TO E-BUSINESS

E-commerce is one part of a much larger phenomenon called *e-business*. E-business involves moving key business processes to the Web in order to gain efficiencies, speed, and marketshare. According to any number of studies conducted in the past year, the evolution to e-business will result in a 60 to 80 percent savings in network costs for many corporations as they migrate away from leased transport facilities in favor of public networks that have the ability to emulate the services provided by a private-line connection.

One technology that has begun to facilitate this is the *virtual private network* (VPN), a recently adopted set of technologies that enable a network user to eliminate the cost of leased facilities without sacrificing the safety and security that they provide. VPNs work as follows: Instead of leasing or owning dedicated facilities between enterprise locations as a way to ensure communication privacy, service providers offer VPN service as a secure alternative. Rather than across dedicated facilities, user traffic is carried across an IP-based public network with traffic from many other users. Each flow of data, however, is isolated from all others through the use of secure protocols that prevent one user from accessing the traffic of another, thus ensuring privacy and security.

This network model has several advantages for both customer and service provider alike. The customer is granted a secure network that guarantees the privacy of transmitted data, yet is not saddled with the cost or complexity of a dedicated network infrastructure. The service provider has the remarkable advantage of being able to resell the same physical network over and over again to different customers, thus allowing them the ability to garner enhanced revenues from a highly capital-intensive resource. So everybody wins.

In October 2001, service provider Global Crossing announced a suite of global VPN services called SmartRoute and ExpressRoute. SmartRoute offers secure IP encryption, class of service management, and bandwidth management. It is considered a network-based VPN because all the intelligence lies within the Global Crossing network. The service offers support for *frame relay* (FR), private line, and *asynchronous transfer mode* (ATM), and it enables a seamless migration path for customers who want to evolve to an IP infrastructure. ExpressRoute, on the other hand, relies on *multiprotocol label switching* (MPLS) to create dedicated paths through the network. Both services are targeted at customers with high-bandwidth requirements.

The retail industry has jumped on the e-business bandwagon with gusto as well. Consider the following scenario: A large retailer with hundreds of outlet stores and many product suppliers implements e-business processes to make their business more efficient. As part of the process, they use decremental inventory control systems to track the on-hand inventory of every store in their chain in real time. Several times a day, inventory information is transmitted to headquarters, where the data is archived and used for sales report generation.

To ensure that their suppliers can anticipate demand and respond quickly to inventory requests, the retailer's internal computer systems are logically connected to those of each major supplier. When a product from a particular supplier in any store hits a low watermark in the inventory system, an order is automatically transmitted from the retailer to the supplier. The supplier responds by sending a confirmation, an invoice to the retailer, and a delivery order to the shipping department with instructions to add that particular product to the outgoing shipment for the store. Meanwhile, the retailer's backroom systems automatically pay the bill using an accepted online payment protocol. Note that the only human involved in this process is the person who drives the forklift on the loading dock and loads the product on the truck to be delivered. Perhaps Warren Bennis, Professor Emeritus at the University of Southern California, was right when he proclaimed that "the business of the future will be run by a person and a dog. The person will be there to feed the dog; the dog will be there to make sure that the person doesn't touch anything."

Some retailers are even more innovative. One major chain combines sales information from each store with local data feeds from the National Weather Service. By passing the sales data and weather information through the digital equivalent of a food processor, they can derive accurate predictive algorithms that enable them to anticipate weather changes and therefore ensure that they have the right products on hand when the weather changes. This takes just-in-time inventory to a whole new level!

And the changes continue: As soon as one form of e-commerce becomes widely accepted, a new one pops up to challenge consumers. The latest of these is called t-commerce, which consists of transactions conducted over interactive television. According to a new study conducted by Jupiter Media Metrix, t-commerce will become an important vehicle for e-commerce transactions in the next few years, most of which will be made using a TV remote control rather than by viewers calling a toll-free number to conduct a purchase. In fact, the primary driver behind digital TV shopping will be a shift away from purchases made over the phone because of the relative simplicity of *point-and-shoot* purchasing.

The parent companies of the major TV shopping channels such as the Home Shopping Network and QVC are poised to reap enormous profits from t-commerce if they can make it a successful transaction model. According to Jupiter, interactive TV shopping programs will pull in $3.4 billion by 2005— income that will be further enhanced by cost savings brought about by the elimination of call centers.

CORPORATE EVOLUTION

To understand the behavior of the modern corporation, it is useful to examine how it evolved from its industrial roots and how the defining characteristics have changed. These include the following:

- The management model that guides the corporation
- The internal corporate structure
- The relationship between the corporation and its peers and competitors
- The nature of the value chain
- The customer's perception of the corporation and its role
- The definition of value in the corporation

Modern corporations have their roots in the industrial age, when corporations relied on the mass production of a commodity to achieve market supremacy. These corporations were largely supplier-driven, because their ability to create products was based entirely on the availability of some scarce physical resource. Furthermore, these industries tended to be vertical in nature, creating a single homogeneous product that was in widespread demand across a variety of *other* vertical industries.

Consider the steel business, for example. During the industrial age, steel was king. The industry, however, was dependent upon a steady supply of ore and personnel, and as long as there was an abundance of both, it prospered. Unfortunately, operational inefficiencies and personnel costs did the American steel industry in as offshore operations figured out ways to do the same job with less people and therefore at a significantly lower cost. Today, there is still a steel industry in the United States, but it is a different business from the one it replaced, manufacturing small-volume lots of highly specialized, custom-designed items. The focus is on customization and service, not product volume.

Businesses can be characterized by two major indicative characteristics. One is the degree to which they innovatively create value in their product to distinguish it from their competitors'. The other is the nature of the resources they use to create the products they sell. If the corporation analyzes their market well and converts the raw material of their business into innovative products, they will succeed and prosper.

Let's consider the business characteristics we discussed earlier as they relate to the industrial age corporation. We identified six of them:

- *The management model that guided the corporation* was a top-down, hierarchical, and rigid structure. All authority was concentrated in the upper tier of the management hierarchy; knowledge was highly compartmentalized.

- *The internal corporate structure* was equally rigid, shaped like a pyramid with multiple levels of management that expanded in number as the levels descended.

- *The relationship between the corporation and its peers and competitors* was somewhat belligerent. Because they sold commodities, price was the only differentiator they typically had. Therefore, they aggressively competed with each other, sometimes in less than business-like ways.

- *The nature of the value chain* was absolutely linear. On one end of the business, raw materials were delivered in trucks; at each progressive step of the manufacturing process, an incremental bit of value was added; finally, at the end of the process, steel popped out. This is an example of the classical fishbone diagram that illustrates the linear nature of legacy industries.

• *The customer's perception of the corporation and its role* was one of awe. These corporations were all-powerful, and in most cases dictated terms to the customer.

• *The definition of value in the corporation* was based on the degree to which they could hold sway over their suppliers, and the degree to which they could outproduce the competition. The name of the game was volume.

As corporations evolved in the 1940s, 1950s, and 1960s, things changed, but interestingly enough many of the afore-mentioned business characteristics did not. Vertical industries evolved (the so-called smokestacks or stovepipes), but many of them, including the telephone companies, continued to rely on the management models of the industrial age. They used a divide and conquer approach, believing that the best way to ensure that a complex task would be completed was to break the task into myriad subtasks, and then to compartmentalize the knowledge required to complete each task.[1] In this model, workers were directed to become highly specialized and to focus on their particular subtask, and not to be concerned with the greater task at hand. If the process required more speed, the corporation simply threw more resources at it—in the form of people, which constituted an inexhaustible resource in many corporations. This model continued to be applied until well into the 1980s in many corporations, including telecommunications.

Today the model is quite different, with the principal goals being to improve operational efficiencies wherever possible and to greatly reduce costs. Furthermore, the six managerial characteristics have changed dramatically. The top-down, hierarchical, centrally controlled corporation has been replaced with a flatter organizational structure, within which is distributed

---

[1]Consider the role of the *Bell System Practices*, or BSPs. These books documented every possible task that could ever be undertaken in a telephone company, broken down to an incredibly detailed degree.

decision-making power at all levels. Instead of adversarial relationships with their competitors, many corporations have crafted cooperative relationships that ensure the survival of all players. The term *coopetition*, which describes this cooperative competition concept, came out of this evolving relationship.

The value chain has changed as well. Instead of being linear and stately, it has become completely nonlinear and moves rather fast. *Value chains* have become *hyperchains*, and customers now perceive the corporation as a peer that is involved as a partner in the success of their business. Furthermore, there has now been defined a value chain of information that does not deal with the processes involved in the manufacture of a finished product, but with the development of knowledge necessary to ensure that it is the *right* finished product.

Perhaps the single most important change is the manner in which corporations perceive value. Industrial age corporations measured value according to their ability to produce massive quantities of products. Internet age corporations measure value according to how well they use the knowledge they acquire about their customers, suppliers, and competitors to create advantageous positions for themselves and how they use that position to provide superior customer service.

In the Internet-age corporation, individuals work in specialized teams, rather than on their own. Those teams may include members from various organizations within the company, from other companies, or even from the customer. Today, for example, it is not uncommon for an ILEC sales team and an equipment manufacturer sales team to jointly call on a customer. As a result of this collaborative-teaming approach, the focus is on the greater task, rather than on the subtask. The team becomes much more future-focused, significantly more responsive to customer requests, and therefore more competitively positioned.

In these corporations, competition exists *not* between individual corporations, but rather between *clusters of companies*. It becomes a matter of survival for businesses to form tight relationships with consumers, suppliers, retailers, distributors, and competitors.

THE IMPORTANCE OF
CORPORATE KNOWLEDGE

In the new marketplace, knowledge-based corporations rise to the top of the competitive heap because they have learned that knowledge about the customer is the single most important resource they have. It isn't enough to have the best router, the most bandwidth, or the most survivable network. Unless the *ore* is manipulated in some way to make it more targeted at each customer's business challenges, it's just more ore.

In a prior life (that is, before telecommunications), I was a scuba diving instructor. One of the skills that all students are required to learn is buddy breathing, the process by which two divers share one regulator. Buddy breathing is used whenever (1) a diver has an out-of-air emergency and (2) there is no clear path to the surface, which might happen if the diver were to run out of air while inside a shipwreck, in a cave, or under a kelp canopy. It turns out that buddy breathing actually has a lot to do with the proper use of corporate knowledge, so bear with me. Besides, it's an entertaining break from technology.

Buddy breathing is an inherently difficult exercise. If a diver finds that he is out of air, he first has to find his buddy, not typically an easy thing to do. Next, he must communicate to his buddy that he is out of air, a process that is accomplished through hand signals that are not intuitive (What? You found a treasure ship?). Once the buddy understands the plight the other diver is in, the two must position themselves to begin buddy breathing. Because the second stage of the regulator (the mouthpiece) always comes over the right shoulder, the person requesting the air must be on the donor's left side. Once they have managed to get into the appropriate positions, they can begin to share the donor's air by passing the second stage back and forth.

This, of course, is where the whole thing usually comes unraveled. By the time the diver realizes that he is out of air, finds his buddy, communicates that he needs to buddy breathe, gets into the right position, and finally starts sharing air, several minutes have passed, after which time the whole process becomes somewhat academic. Even if the process happens without a hitch, the donor naturally assumes that the person requesting the air *is going to give the regulator back*. Because of these minor problems, I always told my students that in the unlikely event that they find themselves unable to breathe because of an equipment problem or because they ran out of air, they should not swim up to me and try to communicate with me using hand signals. The only hand signals I wanted to see, I told them, were those used to pull the regulator out of my mouth and place it in theirs. I'll understand, I told them; I have another one.

So, what does this have to do with corporate knowledge? A lot, as analogies go. A diver should never, under *any* circumstances, run out of air because hanging on the diver's left side is a pressure gauge that indicates, in real time, exactly how much air remains in the tank. *The pressure gauge, however, only works if you look at it occasionally.*

Every night, corporations carefully make copies of their transaction databases. They store the tapes in sealed cases that are picked up the next day by an archival storage company that transports the tapes to a sealed vault where they will reside forever. They are retrieved should a database recovery be required due to a disk failure, but otherwise the information on those tapes stays archived.

Those tapes are like the diver's pressure gauge. They contain invaluable information about customer buying patterns, product return information, and other data that can be converted to a competitive advantage. However, unless the corporation distills that information out of their databases, it is useless to them.

ENTERPRISE RESOURCE PLANNING (ERP)

The need to collect, analyze, and respond to knowledge about customers is of paramount importance today because, if executed properly, knowledge management can become the most critical competitive advantage a company has. This process of collecting the data, storing it so that it can be analyzed, and making decisions based on the knowledge it provides falls under a general family of processes called *enterprise resource planning* (ERP).

ERP is best defined as a corporate planning and internal communications system that widely affects corporate resources. ERP systems are designed to address such functions as planning, transaction processing, accounting, finance, logistics, inventory management, sales order fulfillment, human resources operations and tracking, payroll, and CRM. It serves as the umbrella function for a variety of closely integrated functions that are characteristic of the knowledge-based corporation.

ERP's functional ancestor was an application called *materials requirement planning* (MRP), developed in the 1970s to assist manufacturing companies with the difficult task of managing their production processes and natural resources procurement. MRP systems generated production schedules based on currently available raw materials, and they alerted management when raw materials needed to be restocked.

Over time, MRP evolved into *manufacturing resource planning,* or MRP II. In addition to the tracking and alerting functions of MRP, MRP II enabled production managers to create what-if scenarios that helped them plan for unprecedented events such as a shortage of raw materials or a late start on another project, the delayed output of which would affect the project at hand.

ERP takes the evolution to the next level by integrating financial and human resources concerns into the solution. Unlike MRP and MRP II, which originated in the production industry, ERP focuses more on the business side of the enterprise, concerning itself with the internal operational activities of the company.

The ERP Process

The process by which ERP yields business intelligence folds nicely into the overall convergence hierarchy. During a typical business day, interactions with customers yield enormous quantities of data that are stored in corporate databases. The data might include records of sales, product returns, service order processing, repair data, notes from meetings with customers, competitive intelligence, supplier information, and so on, all the result of normal business activities.

ERP (see Figure 2-1) defines an umbrella under which are found a number of functions, including data warehousing, data mining, knowledge management, and CRM. It defines the process by which corporate data is machined into a finely honed competitive advantage.

That data are typically archived as a database (or multiple databases) in a data warehouse of some sort, typically a disk farm behind a large processor in a data center. At this point, the data are exactly that—an unstructured collection of business records, stored in a format that is easily digested by a computer and that does not yet have a great deal of strategic value. In order to gain value, the data must somehow be manipulated into *information*. Information is defined as a collection of facts or data points that have value to a user.

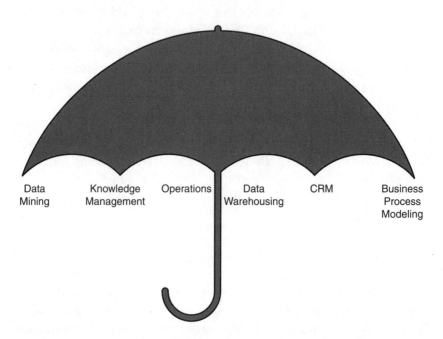

FIGURE 2-1   The Enterprise Resource Planning (ERP) umbrella.

The raw material of information is typically accumulated using a technique called *data warehousing*. Data warehouses store data, but have very little to do with the process of converting it to information. Data warehouses make the data more available to the applications that will manipulate it, but do not take part in the process of converting the data into business intelligence. In fact, according to Brio Corporation, corporate information systems and the user who interacts with them form an *information supply chain* that includes the raw material (data), the distribution systems themselves (the data warehouse and corporate network), the manufacturers or producers (the information technology [IT] organization), and the consumers (end users).

DATA MINING

To convert data into information, a process has been created called *data mining*. Data mining is the technique of identifying, examining, and modeling corporate data to identify behavior patterns that can be used to gain a competitive business advantage. The Gartner Group defines data mining as "the process of discovering meaningful new correlations, patterns and trends by sifting through large amounts of data stored in repositories, using pattern recognition technologies as well as statistical and mathematical techniques." Aaron Zornes of the META Group defines it as "a knowledge discovery process of extracting previously unknown, actionable information from very large databases."

Data mining came about because of the recognized need to manipulate corporate databases in order to extract meaningful information that supports corporate decision makers. The process produces information that can be converted into *knowledge,* which is nothing more than a sense of familiarity or understanding that comes from experience or study. Knowledge yields competitive advantages.

Data mining relies on sophisticated analytical techniques such as artificially intelligent filters, neural networks, decision trees, and analysis tools that can be used to build a model of the business environment based on data collected from multiple sources. It yields patterns that can be used to identify and go after new or not-yet-emerged business opportunities.

Corporations that implement a data-mining application do so for many of the same reasons, including the following:

- **Customer and market activity analysis**   By analyzing the buying and selling patterns of a large population of customers, corporations can identify such factors as what they are buying (or returning), why they are buying (or returning), whether or not their purchases are linked to other purchases or market activities, and a host of other factors. One major retailer claims that their data-mining techniques are so indicative of customer behavior that, given certain conditions in the marketplace that they have the ability to predict, they can dramatically change the sales of tennis racquets by lowering the price of tennis balls as little as a penny.

- **Customer retention and acquisition**   By tracking what customers are doing and identifying the reasons behind their activities, companies can identify the factors that cause customers to stay or go. Domino's, for example, carefully tracks customer purchases, relying on caller ID information to build a database of pizza preferences, purchase frequency, and the like. Customers enjoy the personalized service they get from the company when they place orders. Similarly, Domino's can identify customers who ordered once and never ordered again, and send them promotional material in an attempt to regain them as a customer.

- **Product cross-selling and upgrading**   Many products are complementary, and customers often appreciate being told about upgrades to the product they currently have or are considering buying. Amazon.com is well known for providing this service to their customers. When patrons make a book purchase online, they often receive an e-mail several days later from Amazon.com, telling them that the company has detected an interesting trend: People who bought the book that the customer recently purchased also bought the following books (list included) in greater than coincidental numbers. They then make it inordinately easy for the customer to *click here* to purchase the other book or books. This is a very good example of the benefit of data mining. This so-called shopping cart model has become the standard for online purchases.

- **Theft and fraud detection**   Telecommunications providers and credit card companies may use data mining as a way to detect fraud and assess the effectiveness of advertising. For example, American Express and AT&T will routinely call a customer if they detect unusual or extraordinary usage of a credit or calling card, based on the customer's known calling patterns. The same information is often used to match a customer to a custom-calling plan.

A common theme runs through all these subtasks. Data mining identifies customer activity patterns that help companies understand the forces that motivate those customers. The same data helps the company understand the vagaries of their customers' behavior to anticipate resource demand, increase new customer acquisition, and reduce the customer attrition rate. These techniques are becoming more and more mainstream. The Gartner Group predicts that the use of data mining for targeted marketing activities will increase from less than 5 percent to more than 80 percent over the course of the next 10 years. A number of factors have emerged that have made the mainstreaming of data mining possible, including improved access to corporate data through extensive networking and more powerful, capable processors, as well as statistical analysis tools that are accessible by nonstatisticians because of *graphical user interfaces* (GUIs) and other enhanced features.

A significant number of companies have arisen in the data-mining space, creating software applications that facilitate the collection and analysis of corporate data. One challenge that has arisen, however, is the fact that many data-mining applications suffer from integration difficulties, don't scale well in large (or growing) systems, or rely on a limited number of statistical analysis techniques. Consequently, users spend far too much time manipulating the data and far too little time analyzing it.

Another factor that often complicates data-mining activities is the corporate perception of what data mining does. Far too often, decision-makers come to believe that data mining alone will yield the knowledge required to make informed corporate decisions, a perception that is incorrect. Data mining does not make decisions; people do, based on the *results* of data mining. It is the knowledge and experience of the analysts using data-mining techniques that convert its output into usable information. So Professor Bennis is only partially correct about future businesses being run by a person and a dog.

To work properly, then, data-mining software must rely on multiple analytical techniques that can be combined and automated to accelerate the decision-making process, and that deliver the results in a format that makes the most sense to the user of the application. A number of companies have entered the data-mining game, including SAS Institute, which offers a comprehensive data-mining solution that includes both software and services, and Hewlett-Packard, which has enjoyed a collaborative relationship with SAS since 1997.

Knowledge Management

Data mining yields information that is manipulated through various managerial processes to create knowledge, the golden elixir of customer service. Data mining is an important process, but the winners in the customer service game are those capable of collecting, storing, and manipulating knowledge to create indicators for action. Ideally, those indicators are shared within the company among all personnel who have the ability to use them to bring about change. According to the *Mercer Marketplace 2000 Survey*, 80 percent of those responding believe that the process of transforming information into knowledge is the second most important source of competitive advantage, after CRM.

The collection of business intelligence is a significant component of knowledge management and yields a variety of benefits, including improvements in personnel and operational efficiencies, enhanced and flexible decision-making capabilities, more effective responses to market movement, and the delivery of innovative products and services. The process also adds certain less-tangible advantages, including a greater familiarity with the customer's business processes, as well as a better understanding of the service provider on the part of the customer. By combining corporate goals with a solid technology base, business processes can be made significantly more efficient. Finally, the global knowledge that results from knowledge management yields a more Zen-like understanding of business operations, because of the wide perspective it provides.

OBSTACLES TO EFFECTIVE KNOWLEDGE MANAGEMENT

Knowledge, while difficult to quantify and even more difficult to manage, is a strategic corporate asset. However, because it largely exists in the heads of the people who create it, an infrastructure must be designed that knowledge can be stored and maintained within, and be archived in a way that makes it possible to deliver knowledge to the right people at the right time, always in the most effective format for each user.

The challenge is that knowledge exists in enormous volumes and grows according to Metcalfe's Law. Bob Metcalfe, cofounder of 3Com Corporation and coinventor of Ethernet, postulated that the value of information contained in a network of servers increases as a function of the square of the number of servers attached to the network. In other words, if a network grows from one server to eight, the value of the information contained within those networked servers increases 64 times over, not eight. Clearly, the knowledge contained in the heads of a corporation's employees multiplies in the same fashion, increasing its value exponentially.

There are, however, practical limitations to this observation. According to Daniel Tkach of IBM, the maximum size of a company where people actually know each other and who also have a realistic and dependable understanding of the collective corporate knowledge, is somewhere between 200 and 300 people. As the globalization of modern corporations and the markets they serve continues, how can a corporation reasonably expect to remain aware of the knowledge they possess, when that knowledge is scattered to the four corners of the planet? Studies have indicated that managers glean two-thirds of the information they require from meetings with other personnel; only one-third comes from documents and other nonhuman sources. Clearly, a need exists for a knowledge management technique that employees can use to store what they know so that the knowledge can be made available to everyone.

According to a study conducted by Laura Empson, a recognized authority on knowledge management, 78 percent of major U.S. corporations indicated their intent to implement a knowledge management infrastructure. However, because the definition of knowledge management is still somewhat unclear, companies should spend a considerable amount of time thinking about what it means to manage knowledge and what it will take to do so. Empson observes that knowledge includes experience, judgment, intuition, and values, none of which are easily codifiable.

What must corporations do, then, to ensure that they put into place an adequate knowledge management infrastructure? In theory, the process is simple, but the execution is somewhat more complex. First, they must clearly define the guidelines that will drive the implementation of the knowledge management infrastructure within the enterprise. Second, they must have top-down support for the effort, with adequate explanations of the reasons behind the addition of the capability. Third, although perhaps most important, they must create within the enterprise a culture that places value on knowledge and recognizes that networked knowledge is far more valuable than standalone knowledge within one person's head. Too many corporations have crafted philosophies based upon the belief that knowledge is power, which causes individuals to hoard what they know in the mistaken belief that to do so creates a position of power for themselves. The single greatest barrier to the successful implementation of a knowledge management infrastructure is the failure of the organization to recognize the value of the new process and accept it. Fourth, the right technological substrate must be selected to ensure that the system not only works, but also works efficiently. Corporations often underestimate the degree to which knowledge management systems can consume computer and network resources. The result is an overtaxed network that delivers marginal service at best. Today a significant number of ERP and knowledge management initiatives fail because corporate IT professionals fail to take into consideration the impact that traffic from these software systems will have on the overall corporate network. The traffic generated from the analysis of customer interactions can be enormous, and if the internal systems are incapable of keeping up, the result will be a failed CRM effort and the degradation of other applications as well.

As corporations get larger, it becomes more and more difficult to manage the processes with which they manage their day-to-day operations. These processes include buying, invoicing, inventory control and management, and any number of other functions. The consulting firm AMR estimates that *maverick buying*—that is, the process of buying resources in a piecemeal fashion by individuals instead of through a central purchasing agency—accounts for as much as a third of their total operating resource expenditures, and it adds a 15 to 27 percent overlay premium on those purchases due to the loss of critical mass from one-off purchasing.

According to studies conducted by Ariba, one of the most successful providers of knowledge and supply-chain management software, corporations often spend as much as a third of their revenues on operating resources—office supplies, IT equipment and services, computers, and maintenance. Each of these resources has associated with it a cost of acquisition that can amount to as much as $150 per transaction when conducted with traditional nonmechanized processes. Systems like those from companies like Ariba can help corporations manage costs by eliminating the manual components in favor of computer and network-driven processes. Originally a maker of procurement software for public online marketplaces, Ariba shifted its strategy first to value-chain management and then to *spend management*, creating software that helps purchasing managers analyze and cut costs.

Ariba's *Operating Resource Management* (ORM) application brings order to the knowledge management chaos through the implementation of an electronic infrastructure that enables corporations to strategically monitor, trend, and manage their operating resources. By taking advantage of technology, ORM lowers transaction costs through the use of such components as e-commerce and decision support system automation. By concentrating on the ORM process, corporations can build strategic relationships with suppliers and use the power of aggregate buying to achieve volume discounts.

Another major player in the knowledge management game is SAS. The company's Collaborative Business Intelligence product includes enterprise reporting, data warehousing, CRM, and data mining. This combination of processes unites numerical business intelligence information with text-based, unstructured content. By bringing the two together, businesses can share, distill, and reuse knowledge that would previously have been difficult to locate. SASs' Collaborative Business Intelligence application offers users the ability to search for information in a variety of ways and an innovative subscription mechanism that enables them to receive unsolicited notification when information changes.

Another player in the business intelligence game is Lotus, currently offering a package that is integrated with the company's Domino messaging and collaboration products. The software includes a knowledge management portal that enables customers to organize specific information according to their individual interests and organizational responsibilities. Users can personalize the portal to provide e-mail, calendar services, discussion areas, and quick access to preferred web sites. The application also includes a powerful and customizable search engine that enables employees to search according to personal search profiles.

Several corporations offer products that are specifically targeted at the relationships between telecommunications providers and customers. Many, for example, offer software and hardware combinations that enable customer network managers to monitor a service provider's FR circuits to determine whether the provider is meeting *service level agreements* (SLAs). Given the growing interest in SLAs in today's telecommunications marketplace, this application and others like it will become extremely important as SLAs become significant competitive advantages for service providers.

AN ASIDE: THE SLA

SLAs have become the defining performance standard in the network and systems game. Their role is to define and quantify the responsibilities of both the service provider that is being measured and the customer that is purchasing the service provider's solution. SLAs include the rights a customer has, the penalties that can be assessed against the service provider for failure to perform according to the SLA contract, and, of course, the nature of the delivered service that must be maintained. Their goal is to ensure performance, uptime, and maintenance of *quality of service* (QoS) in environments where networks and systems are critical to the corporation's ability to deliver services to its customers.

The SLA game has become big business. Consultancy IDC has predicted that the overall market for hosted and managed SLAs will grow from an initial $278 million in 1999 to more than $849 million by 2004.

Industry analysts have defined four primary areas that SLAs cover in the modern service-driven corporation. These include the actual network, the hosting services that might be delivered by the service provider, the applications that the client corporation is dependent upon, and the customer service function, which includes technical support, software update activities, and help desk functions. Each of these four management areas is characterized by a collection of *managed entities*, listed here:

NETWORK SERVICE

- The definition of network infrastructure
- Acceptable levels of data loss and latency
- A level of security and a list of the services, applications, and data that are protected

HOSTED SERVICE

- Server availability
- Server administration and management functions
- Database backup and archive schedules

APPLICATION

- Application-specific measurements, including the number of downloads, web page hit counts, user transaction counts, and so on
- The response time for user requests for information

CUSTOMER SERVICE

- The degree to which service monitoring functions are proactive
- The maximum wait time for calls to the help desk
- Technical support availability and wait time

SUPPLY-CHAIN ISSUES

Many of the concerns addressed by the overall ERP process include supply-chain issues. While a supply chain is often described as the relationship between a raw materials provider, a manufacturer, a wholesaler, a retailer, and a customer, it can actually be viewed more generally than that. The supply chain process begins at the moment a customer places an order, often online. That simple event may kick off a manufacturing order, reserve the necessary raw materials and manufacturing capacity, and create expected delivery reports.

A number of companies focus their efforts on the overall supply chain. Most offer an array of capabilities that includes customer service and support, relationship-building modules, system personalization, brand-building techniques, account management, financial forecasting, product portfolio planning, development scheduling, and product transition planning.

ERP IN THE TELECOMM SPACE

ERP's promise of greatly improved operational efficiencies didn't initially attract much attention from telecom carriers, particularly the ILECs. As regulated carriers with guaranteed rates of return and huge percentages of marketshare, they did not rise quickly to the ERP challenge. Recently, however, their collective interest level has grown, as the capabilities embedded in the ERP suite have become better known.

One factor that has accelerated telecom carriers' interest in ERP is the ongoing stream of mergers and acquisitions within the telecom arena. When mergers occur, a great deal of attention is always focused on their strategic implications, but once the merger has taken place and the excitement has settled down, there is a great deal of work left to do internally. Not only does the merged company now have multiple networks that must be combined, they also have multiple support systems that must be integrated to ensure seamless billing, operations, administration, provisioning, repair, maintenance, and all the other functions necessary to properly operate a large network.

Telcos are inordinately dependent upon their *operations support systems* (OSSs) and are therefore more than a little sensitive to the issues that they face with regard to backroom system integration. One of the reasons they have been somewhat slow to embrace ERP is the fact that in many cases they have waited for ERP vendors to establish relationships with the telcos' OSS providers to ensure that together they address interoperability concerns before proceeding.

Another concern that has slowed ERP implementation among telecom providers is fear of the unknown. ERP is a nontrivial undertaking, and given the tremendous focus that exists today on customer service, service providers are loathe to undertake any activity that could negatively affect their ability to provide the best service possible. For example, ERP applications tend to require significant computing and network resources if they are to run properly. Telcos are therefore reluctant to deploy them for fear that they could have a deleterious impact on service.

Furthermore, ERP implementation is far from being a cookie-cutter exercise. The process is different every time, and surprises are legion. The applications behave differently in different system and application environments and, as a consequence, are somewhat unpredictable. Add to that the fact that even with the best offline testing, a certain amount of mystery always remains with regard to how the system will behave when placed under an online load.

Because of the competitive nature of the marketplace in which telecom providers operate today, and because of the fact that the customer, not the technology, runs the market, service providers are faced with the realization that the customer holds the cards. Although there was once a time when the customer might call the service provider to order a single technology-based product, today the customer is far more likely to come to the service provider with the expectation that they be exactly that—the provider of a full range of services, customized for the customer's particular needs and delivered in the most convenient manner.

Because of their access and transport legacy, incumbent service providers understand technology and what it takes to deploy it. As the market has become more competitive, however, they have had to change the way they view their role in the market. They can no longer rely on technology alone to satisfy the limited needs of their customers because those same customers have realized that technology is not an end in itself—it is an important part of a service package. No longer can the service provider roll out a new technology and expect the market to scurry away and find something to do with it; their focus must change from being technology-centric to service-centric. This is not bad news; the technology that they are so good at providing simply becomes a component of their all-inclusive services bundle.

CUSTOMER RELATIONSHIP MANAGEMENT (CRM)

The final component in the ERP family is CRM. It is the culmination of the ERP process, the distilled vintage of information that results from the analysis of business intelligence, which in turn derives from the data-mining effort.

Many companies are offering CRM software today, including PeopleSoft, IBM, Siebel, Hewlett-Packard, Oracle, and others. They too are going through a merger and acquisition period as they jockey to combine their collective capabilities in a market that has suddenly awakened and is loudly demanding their services. PeopleSoft recently acquired sales force automation specialist Vantive, maker of a front-office software package that provides extremely capable service and support functions. The company recently announced its PeopleSoft 8 CRM Suite, which further integrates all the front and backend software systems that the company designs and manufactures. Most players in the CRM game will admit that the key to success in the CRM space is integration with backend applications such as order management, supply-chain, and financial management systems. IBM and Siebel, both major CRM providers, have a cross-marketing CRM relationship in which Siebel sells IBM's DB2 database and IBM sells Siebel's Relationship product. Similarly, HP and Oracle formed an alliance to jointly develop CRM software in September 1999.

Full-blown acquisitions have been in evidence as well. In 1999, Nortel bought Clarify, Inc., which makes CRM software, while E.piphany bought RightPoint. E.piphany software enables companies to collect, analyze, and respond to customer data, while RightPoint's applications add clickstream tracking, analytics, collaborative filtering, and real-time customer-profiling capabilities.

Telecommunications manufacturers have made acquisitions to add to their own capabilities. In 1999, Cisco formed a relationship with integrator KPMG, while Lucent Technologies acquired INS and, as we noted earlier, Nortel Networks acquired Clarify, Inc., a leading provider of front office, CRM, and e-business software. Lucent has since spun off the bulk of the INS acquisition, and Nortel sold its Clarify division to Amdocs Ltd. for $200 million.

PUTTING IT ALL TOGETHER

If we consider the entire ERP process, a logical flow emerges, shown in Figure 2-2, which begins with normal business interactions with the customer. As the customer makes purchases, queries the company for information, buys additional add-on features, and requests repair or maintenance services, database

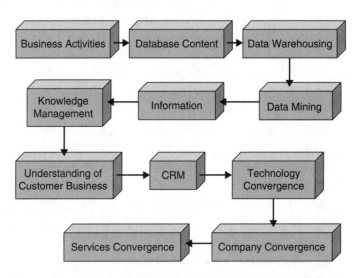

FIGURE 2-2   The ERP process illustrates the corporate decision-making process that leads to proper customer management.

entries accumulate that result in vast stores of uncorrelated data. The data is housed in data warehouses, often accessible through *storage area networks* (SANs), which will be discussed later in this chapter, interconnected via high-speed trunking architectures, such as HyperChannel.

From the data warehouses, corporate data-mining applications retrieve the data and massage it logically to identify trends that help the company understand its customers' behavior, so that it can be anticipated and acted upon before the fact. The data-mining process, then, yields information that now has enhanced value.

The information that is derived from data mining is manipulated in a variety of ways and is combined with other information to create knowledge. Knowledge is created whenever information is mixed with human experience to further enhance its value. The knowledge can then be managed to yield an even more accurate understanding of the customer.

Once customer behavior is understood, strategies can be developed that will help the service provider anticipate what each customer will require in the future. This leads to an enhanced relationship between the company and its customers because the company is no longer in the position of simply responding to customer requests, but in fact can *predict* what the customer will require. This represents CRM at its finest.

Consider this model, then, within the context of the telecommunications service provider space. Once the service provider has developed an enhanced understanding of and a relationship with the customer, and they also know what the customer base will be looking for in the near and long term as far as products and services, the service provider can develop a technology plan that will help to ensure that they have the right technologies in the right place at the right time to satisfy those service requests. Given the accelerated pace of the telecommunications market, however, service providers do not have the time to develop new technologies in-house. Instead, they do the next best thing, which is to go to the market, identify another company that offers the technology they require, and either buy the company or form an exclusive, strategic alliance with them.

The service provider gains the technological capability they require to satisfy their customer, the technology provider gains an instantaneous and substantial piece of marketshare, and the customer's requirements are not only met, but they are anticipated.

This model of providing a full complement of services by a group of companies instead of a single provider is a relatively new phenomenon in business. It is often referred to as the *virtual community* and is one of the most powerful change agents in telecommunications today.

THE BIRTH OF THE VIRTUAL COMMUNITY

As corporations evolve from standalone competitive entities driven by suppliers to knowledge-driven clusters driven by the customer, a number of qualities emerge that characterize all companies. The two most compelling of these characteristics are (1) the degree to which the corporation allows its internal entities to self-manage and (2) the value the corporation adds to the product through its own actions. We will examine each of these.

MANAGERIAL CONTROL

In terms of managerial control, corporations range from being completely hierarchical, with all decision-making power concentrated in the upper managerial echelons, to totally self-managed entities where decisions are made by lower-level managers on a peer-to-peer basis. Hierarchical management certainly works; it maintains tight control over all functions within the corporation such that senior managers have the ability to control the affairs of the business. A hierarchically managed corporation is effective when competitors and customers are also hierarchically managed, but tends to be seen as unresponsive and pedantic by customers or competitors that have adopted the peer-to-peer model. It is difficult to respond quickly to requests for service or to a competitive threat when all deci-sions must pass through a hierarchical managerial bottleneck.

Self-managing entities tend to move quicker, make decisions more effectively, and respond to competitive threats and customer requests more nimbly than their hierarchical associates. Their challenge is one of intracorporate communication. Because they lack centralized authority, communications can suffer between the peer entities. Successful self-managed corporations have real-ized that even though they are structurally flat, ideologically they are hierarchical. Although the corporation may be comprised of a large number of loosely connected business "molecules" that work independently, they all work toward a common goal that is well known by all, such that their collective efforts serve the over-all goals of the corporation (see Figure 2-3).

ADDITION OF VALUE

Value is added to a corporation's products in a variety of ways, ranging from nothing more than a place to buy them (minimal added value) to integration services that help customers posi-tion products and services to their own competitive ends. For example, an ILEC might provide nothing more than access and transport, in which case its added value is minimal. On the other hand, it might provide consulting services that help its

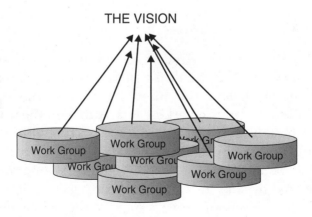

FIGURE 2-3   The virtual hierarchy in the virtual corporation.

customers select the appropriate technology solutions based on what they know about their own customers. In this case, the added value is substantial, and the corporation's value in the eyes of its customers is enhanced.

As companies evolve in the information-driven economy, the degree to which they exhibit these two characteristics allows them to be categorized according to a limited set of behavior models. This model is useful when analyzing companies because the unique combination of managerial independence and embedded value indicates the nature of the company, the services it provides, and the customers it can hope to satisfy. It also serves as a powerful self-analysis tool. If a corporation determines after analyzing itself according to this model that it falls into a less-than-optimal category, it can use the model to determine what it needs to do differently to move itself in the proper direction.

COMPANY CLASSIFICATIONS

As Figure 2-4 shows, corporations fall generally into one of four models: the flea market, the arbitrator, the process improver, or the virtual corporation. All have their place, and in reality most corporations exhibit characteristics of all four, as indicated in the figure, although one is typically dominant.

THE FLEA MARKET MODEL

Most of us have been to a flea market at one time or another, whether while visiting Madrid's Rastro, Buenos Aires' Plaza San Telmo, or one of the hundreds that spring up every Sunday in drive-in movie parking lots. Although these markets appear to be a form of organized chaos, a thread of management runs through them. First, the flea market is comprised of myriad independent, self-managed vendors, selling their merchandise side-by-side in a large, open market structure. And although there is a management overlay, the only added value they bring is the space they provide and the folding tables they

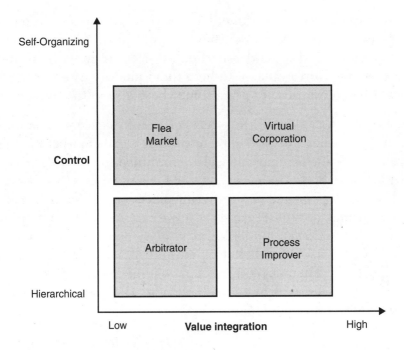

FIGURE 2-4   Corporate models.

rent to the vendors. Otherwise, they are invisible to vendor and customer alike.

In the flea market company model, several characteristics are evident: First of all, the role of buyer and seller is purely serendipitous in the sense that buyers today may well show up next week as sellers themselves. Very little control exists in this model; it is completely self-managed at an organic level. Furthermore, there is very little in the way of organizational trust here. Because of the independence of each of the vendors that make up the overall market, there is minimal coordination, no hierarchical oversight, and therefore minimal trust. *Caveat emptor* is the watchword.

How does market leadership reveal itself in this model? Again, serendipity plays an active role. Market leadership has more to do with being in the right place at the right time, with the right product at the right price. Leadership is fleeting in the flea market; commodities sell well here because they are differentiated on price alone. Unfortunately, from a customer's perspective, the flea market model is not ideal. Although customers may find a good price for the item they wish to buy, there is no effort available to help them identify whether there might be an even better price somewhere else in the market.

Flea Markets in the Telecommunications Market Although it may not be completely obvious, flea market models are at work in (or near) the telecommunications marketplace. Online auction house eBay is a good example; eBay itself provides nothing more than a gathering place for thousands of sellers to display their wares and for buyers to bid on them, hoping to secure the lowest possible price. eBay provides little added value, other than the digital equivalent of a parking lot and a folding table. Their success, of course, indicates that the service they *do* provide is a highly valued one. It is not, however, a service-differentiable organization.

In Summary   Flea market businesses offer little in the way of enhanced services beyond their basic products (by design) and exhibit very little coordinated management of business processes. This is by no means a criticism of businesses that model themselves this way because many of them are quite successful.

The Arbitrator Model

As we move up the organizational food chain, we come to the arbitrator model. Added value is still relatively low, but unlike the flea market model, here we find hierarchical management practices in use. In the arbitrator model, we find businesses that have found a market niche by placing themselves between buyers and sellers, and by serving as an intermediary between them. As a result, there is a recognized market leader because their service, while not highly differentiated, is valuable and necessary.

In many cases, arbitrators are very large and command significant power in the marketplaces in which they operate. As a result, they can often aggregate supply and demand by negotiating favorable prices with suppliers, and then providing comfortable gathering places for customers. In the retail world, companies like Home Depot, Costco, Staples, and Wal-Mart fulfill this role. Because of their buying power, they can create ideal market conditions for buyers and sellers, aggregating supply and demand in the most efficient manner possible. Although the service they provide above and beyond their basic selling role is valuable, it is targeted at the mass market and is therefore not particularly specialized.

Arbitrators also enjoy greater trust than their flea market counterparts, because they often develop and rely on widely known brand loyalty that carries a high degree of trust on the part of the buyer.

ARBITRATORS IN THE TELECOMMUNICATIONS MARKET Arbitrators are the best-known players in the telecommunications arena, including such names as *America Online* (AOL), E*Trade, Prodigy, and other nationally known online service providers. AOL, for example, provides a safe gathering place for buyers of a remarkable variety of products and services, while at the same time providing a gathering place for the sellers themselves. Furthermore, because of their staggering market power, they can dictate rules of behavior to create the marketplace image they desire.

A recent study observed that a remarkably large percentage of AOL subscribers never leave AOL—that is, they never avail themselves of the Internet portal that is available a single click away, an indicator of the degree to which AOL alone satisfies their requirements for an online service provider. As one of the company's supporters recently observed, "Think of AOL as the Love Boat. You pay one price, for which you get everything. It's a totally safe place where you'll never get into any kind of trouble or run across anyone unsavory. You can go anywhere you want, and they have something for literally everyone. The problem is that you can get off the boat when it docks. And if you do, there's no telling what might happen. You could get robbed, or worse. So the best thing to do is stay on the boat!"

IN SUMMARY   Arbitrators sit between buyers and sellers, bringing them together and using their considerable market power to ensure that they can offer a wide variety of products and services for the most competitive price available. They also rely on brand loyalty to ensure that they form long-term relationships with customers.

THE PROCESS IMPROVER MODEL

One level up on the corporate food chain we find the process improvers. Here we find a leader of the pack who differentiates him- or herself by improving the overall process by which his or her business niche operates. Process improvers tend to be higher on the added-value scale, but also tend to be managed in a more top-down fashion. Instead of serving simply as a gathering place or as an aggregator of buyers and sellers, process improvers optimize the value chain to create greater value in the service they provide above and beyond the simple products or services they sell. As a result, the value they add is substantially higher than the two models we have seen previously. Automobile dealers who interview their clients and then tune preset radio stations according to their preferences are exhibiting process improvement characteristics.

PROCESS IMPROVERS IN THE TELECOMMUNICATIONS MARKET Online service providers such as Amazon.com, Dell, and Cisco are examples of process improvers in the telecommunications marketplace. Amazon.com not only sells books and other products, but it also gathers customer preferences, and then upsells its customers according to those preferences, by sending them discount coupons and notifications of other books the company thinks the customer might like based on logged preferences. Dell interviews customers online as part of the presell activity, thus ensuring that they understand exactly what the customer's requirements are before attempting to sell them a machine. Consequently, they are far more likely to satisfy the customer with their first-time purchase and keep them as a long-term customer.

IN SUMMARY   Process improvers optimize the value chain, thus adding value to the products and services they sell by ensuring that doing business with them will result in the most cost-effective and efficient experience for the customer. They lead by being the best at what they do, not necessarily by having the lowest price. They count on the fact that their customer will perceive that there is an inherent value in doing business with the company.

THE VIRTUAL CORPORATION MODEL

The virtual corporation is perhaps the least understood and least visible of all the business models described. Here we find that there may actually be multiple leaders, all providing specialized services to narrowly defined segments of the marketplace. The goal of the virtual corporation is to strive to understand—and perhaps even predict—specific customer challenges and provide optimized, targeted solutions for them. The virtual corporation model exhibits a phenomenon often called *the convergence of money and minds*, characterized by the fact that the market decides where the money should be directed within the industry based on the fact that a particular product or service satisfies a critical demand and therefore deserves funding. Once the market's tractor beam has magnetically directed the money as it sees fit, the talented people will naturally follow the money.

Consider the remarkable success of 3Com's Palm Pilot product. In quick order, the market realized how capable the device was and directed money to its manufacturer by buying Palm Pilots in the tens of thousands. Driven by the success of 3Com, talented people flocked to the corporation in droves to jump on the success bandwagon. In concert with this frenzy, a plethora of related corporations appeared like mushrooms on a summer lawn, selling third-party software and adjunct devices. Furthermore, competitors such as Mindspring and the Palm PC market evolved, all drawn by the magnetism of money and minds.

Corporations that exhibit the characteristics of a virtual corporation add a great deal of value to the basic products and services that they sell. They tend to be peer-to-peer management entities, relying on the abilities of individual work groups to perform necessary tasks according to some overarching corporate vision. These work groups may form unpredictably and typically do not exist for long periods of time. They remain together long enough to accomplish a task, and then dissolve and go on to other mandates. One interesting point about these virtual entities: They tend to cross organizational boundaries, pulling in members from the ranks of other corporate organizations, other companies, and even from customers to ensure that they have the appropriate ability at hand to accomplish the task.

VIRTUAL CORPORATIONS IN THE TELECOMMUNICATIONS MARKET
A number of good examples of the virtual corporation model exist. These include the *Joint Electronic Payment Initiative* (JEPI) and the *Secure Electronic Transaction Set* (SET), both made up of companies dedicated to the creation of a widely accepted set of online payment protocols. Other examples would be the Java Alliance, which is comprised of Sun, IBM, Oracle, and Netscape; Microsoft and Intel's Wintel Alliance, designed to bolster support for the PC domain; and the Universal ADSL Working Group created by Microsoft, Compaq, and Intel, designed to bolster support for *asymmetric digital subscriber line* (ADSL). Each of these collectives has a particular goal that they want to achieve, and in some cases the members of each compete with each other.

These are, of course, pre-packaged in many ways, having come about because of specific long-term challenges that needed to be addressed. However, when examining the incumbent players in telecommunications today, it is an interesting exercise to determine where they each fall and to ask where they *should* fall if they had the ability to change. If we examine the makeup of the market, we find nine major groups: the ILECs, the CLECs, the *Interexchange Carriers* (IECs), the ISPs and portals, the cable providers, the equipment manufacturers, and everybody else. Each has a well-planned strategy for maintaining and growing marketshare, but the manner in which they do so varies from company to company. In the section that follows, we will examine these industry components as generic groups, but will then look at specific companies to understand how they view their own roles as the marketplace evolves.

THE ILECS

By and large, ILECs tend to be relatively homogeneous in terms of the products and services they provide. Their strengths lie in the access and transport business at which they excel; however, because of the commoditization of this business, their ability to maintain marketshare is diminishing. For the most part, they have grown by acquiring more of the same. Witness Bell Atlantic's acquisitions of NYNEX and GTE to begin the formation of Verizon, or SBC's acquisitions of Pacific Bell, Nevada Bell, Ameritech, and SNET. They have expanded their footprint, but have not done much to diversify their product and service offerings.

In fairness, there are good reasons for this strategy; the ILECs have realized that with long-distance relief pending, they must create a wide area presence for themselves. Their larger business customers are not necessarily local companies. Although they may have a local presence, they tend to be national or even global. If the ILECs are to become full-service *service providers*, they must be able to serve those customers on an end-to-end basis, thus eliminating the need for intermediaries. Without a wide area data network, they cannot accomplish this.

The fact is, the market is the ILECs' to lose. Many analysts believe that customers will buy all services from the local service provider if it has the ability to provision them. The holder of the access lines rules. Consequently, much of the company convergence activity of late has revolved around the acquisition of access lines. Consider Qwest's acquisition of USWest, or Global Crossing's acquisition of Frontier. On a slightly different level, AT&T's acquisition of TCI was clearly a gambit for local loops, and although the model has not proven to be as successful as they might have liked, more will follow.

Today, ILECs fall largely in the flea market model. The product they sell is fast becoming a commodity, and although they are good at what they do, they do not add a great deal of additional value above and beyond the access and transport services they provide. The flea market model tends to reflect more of a peer-to-peer management style, but the ILECs exhibit more of a hierarchical management model.

So, are ILECs a dying breed? Will they be brought down by the smaller, more nimble CLECs that are nibbling away at their long-standing customer bases? There is no question that they face some challenges. Their networks were designed around the idea that they would control 100 percent of the market and are therefore not the most cost-effective resource in an open and competitive market. Other models are far more cost-effective than the ILECs' circuit-switched infrastructures. As a result, the ILECs are reinventing themselves one piece at a time and are, of course, expanding their market presence in a variety of ways. As we noted at the beginning of the section, however, regulatory forces and the incumbent power of the ILECs give them an enormous advantage in their market, an advantage that they will not give up easily.

THE CLECS

The CLECs are similar to the ILECs in that they sell a commodity. They differ from the ILECs, however, because they can be selective; they tend to sell the more lucrative products and avoid the markets and services that don't enjoy high returns. For example, many CLECs focus on residential voice customers, while others exclusively pursue business customers. All CLECs are not created equal; their business strategies and business plans for carrying out those strategies and satisfying customers vary dramatically from company to company. Many of the larger ones have disappeared, such as ICG, Teligent, and Winstar. Many are, however, good performers within specific bounds. In rural areas, they have in many cases carved out service niches for themselves within which they offer a very high level of customer service.

CLECs face significant obstacles by virtue of the fact that they are CLECs. As alternatives to the ILECs, they rely on interconnection agreements with ILECs because CLECs must have a collocation presence within the ILECs' central offices to provide service. The 1996 Telecommunications Reform Act mandates that before the ILECs will be allowed into the long-distance market, they must demonstrate that they have opened their local market to competitors, allowing equal access to unbundled facilities such as local loops and certain services. CLECs often complain that while the ILECs have agreed to the stipulations, they are not particularly quick to respond to CLEC requests for interconnection services and therefore have the ability to exert some control on the pace at which CLECs can enter their markets. There are, of course, some checks and balances in place, such as Section 251 of the 1996 Communications Act. This component of the law requires that ILECs sell circuits, facilities, and services to their competitors that are "at least equal in quality to that provided by the local exchange carrier to itself or to any subsidiary, affiliate or any other party to which the carrier provides interconnection."

CLECs are similar to the ILECs; they fall somewhere between the flea market and arbitrator business models, selling commodity access and transport while trying very hard to add services that yield a competitive advantage over the ILECs. This is important; the ILECs control 90 percent of the access lines in the United States, so CLECs face a significant challenge. Many CLECs claim to be able to offer better, more customized service than their ILEC competitors. Most customers agree that the technology products sold by the ILECs and the CLECs are identical. The difference, they claim, is the way they deal with their customers. CLECs believe themselves to be more customer focused, claiming that the ILECs are still plagued by legacy monopoly mentality. Whatever the case, some CLECs have initiated differentiation programs to help them garner the favor of customers, such as payback plans for downtime, online real-time usage reports, and negotiated SLAs.

Ultimately, the success of the CLECs relies on three critical success factors: Local number portability must be viable, functional, and available; operations' support standards must be in place and accepted; and discounts for unbundled network elements from ILECs must be on the order of 50 percent.

THE IECS

The IECs face the greatest challenge of all the players, but are also the companies demonstrating the most innovative behavior in the face of adversity. With companies like Qwest and Level 3 building massively *overcapacitized* fiber networks, bandwidth is becoming so inexpensive and so universally available that it is evolving to a true commodity. The margins on it, therefore, are dropping rapidly. Furthermore, the number of companies that have entered the long-distance market has grown, as has their diversity. Although the legacy players (AT&T, Sprint, and MCI) continue to hold the bulk of the market, a collection of power companies, satellite providers, and *bandwidth barons* have entered the game and are seizing significant pieces of market-share from the incumbents.

In response, the IECs are fighting back by diversifying. All have entered the ISP game, offering Internet access across their backbones at competitive prices. They have also bought or built local twisted pair access infrastructures, cable companies, satellite companies, and a host of others. Consider AT&T as an example. Beginning with their long lines division, the company has grown into a multifaceted powerhouse that acquired IBM Global Networks, TCI, Teleport, Excite, and @Home, to name a few. They are a long-distance, local, wireless, ISP, portal, cable company and have aggressive plans to be a true, full-service telecommunications provider as they flesh out their strategy for market positioning. However, the company has recently reduced its cable holdings (Excite@home is gone, and TCI has been absorbed as part of AT&T Broadband) and is focusing instead on a limited number of capabilities.

The IECs, like the ILECs and CLECs, fall somewhere between the flea market and arbitrator models, although they are edging toward the arbitrator side of the equation as they diversify their holdings. Some of them are becoming virtual corporations as they diversify.

THE ISPS AND PORTALS

The ISPs represent something of a mixed bag of business models. The small ISPs that provide nothing more than IP access to the Internet fall solidly into the flea market model, inasmuch as they provide a nondifferentiated service among a host of other companies doing the same thing. The larger ISPs, however, have differentiated their service offerings to varying degrees and have formed alliances with other players in rather innovative ways. They have become portal providers as well, offering one-stop shopping/searching/gathering capabilities to their customers.

AOL is the most visible example of a highly differentiated ISP. First of all, the company's service is targeted at the average consumer, not the *propellerheads* that many ISPs cater to. Second, the company is diverse. It owns Time Warner (which includes *Time, People, Entertainment Weekly, Fortune, Sports Illustrated*, Warner Bros. Studios, Warner Music Group, Time Warner Cable, CNN, HBO, and others), CompuServe Computer Online Services, Netscape, and ICQ. AOL also has strategic alliances with a number of key players, including Sun Microsystems, Gateway, and TiVo. It has marketing agreements with SBC and Bell Atlantic, and hundreds of agreements with content providers to serve as their sole hosting service. To their credit, AOL has 33 million customers and shows no signs of slowing down.

THE APPLICATION SERVICE PROVIDERS (ASPS)

One of the newer service segments to arrive on the scene is the *application service provider* (ASP). ASPs deliver and manage applications and computer services from remote data centers to customers, either via the Internet or via a private network. Their goal is to offer a cost-effective solution to the cost and complexity of systems management, acquisition, and ownership, including the reduction of up-front capital expenses, implementation challenges, and a continuing need for maintenance and customization.

ASPs offer customers a cost-effective alternative to the procurement and operation of complex software systems. Customers can control the cost of technology ownership through scheduled payments, thus better managing the overall cost of service management. Because the IT and data processing functions are performed at a remote site by a third party, the client can focus on core functionality rather than IT. An ASP also has numerous benefits, including enhanced speed to market, improved internal and external performance, increased financial flexibility, and reduced overall risk.

An ASP relationship can be cost-effective, and it simplifies system management and support. ASPs deliver applications and services from remote data centers, and primarily target small and mid-sized organizations. This is not to say that no market exists for their services and capabilities in large enterprise corporations, but the charm and attraction of leased applications are particularly appealing to smaller entities, because they can deploy enterprise applications that would otherwise involve massive investments in software, deployment time, and IT personnel.

IT personnel are beginning to buy into the ASP concept as a way of ensuring the availability of secure, business-critical applications, and ASPs are among the first to loudly support SLAs as part of their offered services.

THE MANAGED SERVICE PROVIDERS (MSPS)

Managed service providers (MSPs) are yet another new arrival in the service provisioning space. They offer such services as remote monitoring and the administration of computer systems, while allowing customers the security and control that comes from keeping their hardware and systems at their own sites. This is different from the ASPs described earlier who host applications and hardware remotely.

MSP services are best suited for small and medium companies that can't afford or don't have the trained IT staff required to support a growing and increasingly complex IT infrastructure. Customers have the option of selecting from various QoS levels that are appropriate to their particular requirements, ranging from best-effort service for such applications as e-mail and Web access to highly reliable service for so-called mission-critical applications.

Among the preferred customer targets for MSPs are e-businesses. These companies are often strong in business management skills, but lack IT experience, an equally critical component of a business plan.

The benefits of a relationship with an MSP can be quite strong. For new companies, the ability to purchase an entire, turnkey, off-the-shelf IT department offers speed and flexibility, instantaneous functionality, peace of mind, and zero recruitment ramp-up time.

MSPs offer a variety of services. They typically manage and monitor customer networks using secure Internet connections, and they provide performance monitoring as well as system and database administration. Some MSPs also host applications, which has a tendency to confuse some customers who are trying to decide between the services offered by ASPs and MSPs. Generally speaking, MSPs go after the enterprise management demand for services, while ASPs provide offsite implementation.

So, where do these companies stand? The pure, low-end ISPs are clearly flea market companies, offering very little value integration. The larger ISPs, ASPs, and MSPs, however, that host diverse content and provide portal services fall more into the arbitrator model. They aggregate buyers and sellers, providing a safe, comfortable place for them to meet and do business. In some cases, they may exhibit characteristics of the process improver, creating order from chaos through business process enhancement.

THE CABLE PROVIDERS

Similar to the ILECs, cable companies provide a largely homogeneous service that is under attack by a variety of alternative providers. Although they are striving to provide diverse services, the challenge to do so is daunting. Most of them have converted their networks to all-digital architectures and offer two-way data service through installed cable modems, thus expanding their service offering and entering new lines of business. They fully intend in many cases to offer online content at some point; in the meantime, however, they sell commodities and are therefore flea markets.

THE EQUIPMENT MANUFACTURERS

Faced with serious competition, rapidly advancing new technologies, and demands for service from the marketplace, the major (successful) equipment manufacturers exhibit company convergence at its finest. Consider Cisco, for example: In the last three years, the company has acquired more than 40 companies, all designed to expand Cisco's abilities in the burgeoning telecommunications market. Long known as a data company, Cisco reinvented itself and now offers an impressive line of integrated voice products as well. Nortel Networks and Lucent Technologies have followed similar routes, building products that will satisfy a diverse set of required capabilities. As such, the manufacturers are flea market players, although they exhibit characteristics of all four company types.

EVERYBODY ELSE

So, who falls into this category? Power companies occupy the largest niche, selling bandwidth on their own networks when the price is right. Other companies with rights-of-way such as railroads and pipeline companies are also found here; they are largely based on the flea market model.

AND THE POINT IS . . . ?

Alice: Which way should I go?

Cat: That depends where you are going.

Alice: I don't know where I'm going!

Cat: Then it doesn't matter which way you go.

Lewis Carroll, Through the Looking-Glass

Three basic truths govern the decisions that companies are making today to define themselves within the models described in the last section. The first of these is that access and transport are commodity products that can be bought from a variety of providers at ever-lower rates. Second, if access and transport companies want to continue to maintain an edge in their marketplace, they must climb the services food chain and become more than what they currently are. They must become full-service telecommunications providers because more and more often customers are looking for all services from a single provider. *All services,* incidentally, implies exactly that; the ability to offer a broad spectrum of bandwidth in multiple flavors is not enough. Customers want services, content, Internet access, and more, all delivered properly and billed under a single, itemized invoice. The third of these is an observation about technology evolution: Wireless access is creeping in, and the percentage of fixed wireless local loops is climbing steadily. Some reports claim that as many as 40 percent of today's wire-based local loops will be replaced by wireless in the next few years.

No single company in the telecommunications industry today is capable of accomplishing this service reorientation on their own, yet their very survival is based on their ability to do so. The only solution is to pool resources with other players, creating a business "molecule" that combines the best of multiple service "atoms" to create the combination of strengths and abilities desired by the customer base. There is certainly a need for all forms of business structure, as defined in the previous section, but it should be relatively clear at this point that service providers that have plans to stay in the game *must* take steps to move themselves into the virtual corporation quadrant.

The reasons for making this more are clear. First, they recognize that they are simply not capable of providing the services customers desire today, given their current business model. Second, inasmuch as service is the name of the game, they must form alliances with other corporations so that they can provide a packaged bundle to their customers as a defense mechanism against competitors. Third, if they manage their mergers and alliances properly, customers will be unaware of the fact that in reality they are receiving services from multiple providers because they will do business with a single entity, behind which lies a cloud that disguises the complexity of the corporation.

Companies that elegantly demonstrate this converged mentality are Lucent Technologies, Cisco, and AOL. Lucent Technologies and Cisco are both telecommunications equipment manufacturers and powerful competitors. Lucent is well known for its flagship 5ESS circuit switch, while Cisco is recognized as the premier provider of packet-switching-based products. AOL is an online service provider, moving to recreate itself as a virtual corporation.

COMPANY CONVERGENCE: LUCENT TECHNOLOGIES

Lucent is a classical example of a virtual corporation. Originally founded as Western Electric—the manufacturing arm of the AT&T corporation—Lucent has undergone a number of significant changes over the years. Since being spun off from AT&T in 1996, the company has acquired more than 30 other companies, including

- **Ascend**   ATM core switching
- **Yurie**   ATM edge switching

- **Optimay** *Global System for Mobile Communications* (GSM) call-processing software
- **Stratus** *Signaling System* 7 (SS7) switching
- **Quadritek** IP software
- **Sybarus** Semiconductor design
- **WaveAccess** High-speed wireless Internet access
- **Livingston** Remote access products
- **Prominet** Gigabit Ethernet
- **Nexabit** IP switching and routing
- **Mosaix** Front-end/back-end system integration
- **SpecTran Corporation** Optical fiber
- **International Network Services (INS)** Consulting and systems integration
- **Excel** IP-based programmable switching

If we consider the products represented here, we find that Lucent has done an exceptional job of shoring up the weaknesses in its product line, specifically the data side of the telecommunications house. The 5ESS is a well-known solution for telephony providers, and Lucent has long been recognized as a leader in legacy voice. It has, however, had a name as a data solutions provider until recently, when its acquisitions activities started shortly after being spun off of the AT&T juggernaut. Today the corporation has clearly become a multifaceted player, with diverse capabilities that include

- Circuit-switch capabilities, including signaling
- ATM core and access switch technology
- Fast and Gigabit Ethernet
- Wireless telephony and Internet access
- IP platform software
- IP switches and routers

- Front-end to back-end system integration
- Systems consulting and integration

Clearly, Lucent has realized that it cannot continue to be a traditional provider of circuit-switched technology and has taken aggressive steps to correct their course. They have relied on the adage the quickest way to become a leader is to find a parade and get in front of it. The parade, it seems, is the long string of companies Lucent has either acquired or formed alliances with. Lucent has reinvented itself as a virtual corporation capable of delivering every possible technology solution that a customer might request. In fact, the 7 R/E switch, the evolutionary replacement for the 5E, reflects the company's acquisitions and alliances to date. The new switch boasts components and software from many of the companies that now bear the Lucent logo. Yet when customers buy a 7 R/E, it bears the Lucent brand.

Lucent has been slammed like everyone else in the industry, and unfortunately had the dubious honor of being the bell-wether of what was to come with the so-called meltdown. In the last few months, the corporation has gone through huge staff layoffs, shed product lines, restructured lines of business, and redefined their role in the industry. Today they have a respectable 17 percent of the worldwide voice-switching market, 14 percent of optical transports, 20 percent of the Internet infrastructure, 18 percent of wireless, and 25 percent of *frame relay* (FR) switching.

The company recently underwent a major functional restructuring of their product lines. In press releases leading up to the announcement, Lucent said the global market continues to challenge them, and in response they have realigned themselves into two segments: integrated network solutions and mobility solutions. Under the new vision the company has developed, Lucent provides a vision of service intelligence, a complete portfolio of products, a comprehensive network management suite, a global systems integration force, and high-end, in-house *research and development* (R&D). The company has stated that on a worldwide basis, service provider capital expenditures are returning to sustainable levels and will reach $259 billion for 2002.

Lucent's product line has now been realigned according to their revamped organizational structure and now includes a variety of product portfolios. The Core Optical Portfolio includes the LambdaRouter all-optical switch, the LambdaXstream long-haul and ultra-long-haul transport platform for 10 Gbps and 40 Gbps transports, the LambdaUnite system that bridges or unites 10-Gbps and 40-Gbps traffic across metro networks, and the LambdaManager, a terabit *time division multiplexing* (TDM) system to manage subwavelength or electronic traffic in ring or mesh topologies. The LambdaXstream will use a combination of soliton transport and Raman amplification with programmable add/drop and embedded service intelligence.

The MultiService Core Transport Portfolio includes the TMX 880 Multiservice Xchange switch (January 2002) designed to facilitate the transition from ATM to MPLS core networks, the GX550 multiservice switch with a new processing card that boosts performance by 400 percent for the core or edge, the CBX 500 multiservice switch for edge applications, and the B-STDX FR ATM switch.

The Metro Optical Portfolio includes the Metro DMX *Synchronous Optical Network* (SONET) platform that integrates cross-connect and *add/drop multiplexer* (ADM) functions. It also offers efficient TDM and IP/Ethernet transport as well as a metro *enhanced optical network* (EON) system designed to reduce the cost of *dense wavelength division multiplexing* (DWDM) deployment.

The Edge Access Portfolio includes the APX 8000 MAX TNT with universal port capabilities for dial-up and *voice over IP* (VoIP) services, the PSAX family of ATM access concentrators, the APX 1000 carrier-grade concentrator for backhauling wireless traffic, the Stinger *digital subscriber line access multiplexer* (DSLAM) line for central office and remote terminal installations, and the AnyMedia Access System digital loop carrier.

The IP Services Portfolio includes the Springtide 5000/7000 IP services switch, the Springtide 7000 wireless IP services switch, and the Lucent Softswitch 3.2. They will also offer a range of hardware and software products designed to support circuit-to-packet migration, including a 5ESS Packet Trunk solution, a 5ESS Large Tandem solution, a 7 R/E Packet Tandem solution, and a Softswitch T3 solution.

Finally, the Mobility Portfolio includes the Flexent Base Stations for CDMA 1, CDMA 2000, and *Universal Mobile Telecommunications System* (UMTS) protocol support, as well as radio network controllers for CDMA 2000 and UMTS. Lucent will also offer a SuperHLR for authentication, subscriber profiles, and IP addresses for mobile users, as well as packet data gateways and mobile switching centers based on the 5ESS and Softswitch.

As Lucent practices convergence and redefines its business services model, the company will get out of the fixed wireless business and the cable access market. As part of the product-line refinement process, they have also dropped its Chromatis metro optical product line; the Chromatis products, though very good, were primarily targeted at the CLEC market.

The other area in which Lucent has made a significant effort is network management. The company is redeveloping its network management products into a suite for the management of distinct network elements, service provisioning, and QoS assurance across circuit, packet, optical, and mobile networks, all from a single platform. Known as Navis iOperations, the product combines five distinct operations support systems. These include Navis iProvision, iAssure, and iEngineer for fault, configuration, accounting, performance, and security management of Lucent access, ATM/FR, and optical and IP network elements. Two new modules include the Navis Radio Configuration Manager and VitalEvent event analysis software. The Radio Configuration Manager validates transient and permanent network configurations, sets policies and plans radio frequency capacity, and reconfigures cell sites.

COMPANY CONVERGENCE: CISCO

Lucent, of course, is not alone. Cisco Corporation is another excellent example of the virtual corporation model. Founded in December of 1984 in Menlo Park, California, Cisco now has more than 10,000 employees working in 200 offices in 54 countries. Cisco does not rely on a single technology approach to the market. Their philosophy consists of listening to customer requests, assessing technological alternatives, and giving customers a range of options from which to choose. Their products are designed around internationally recognized standards, although some of their products rely on proprietary solutions designed to satisfy specific problems not adequately addressed by existing standards. In fact, some of Cisco's technologies have *become* industry standards. Cisco was also the first company to deliver multiprotocol routers. This gave them a head start over the competition and allowed them to become the leading supplier of enterprise networking solutions.

To maintain market leadership, the company has rapidly developed and introduced new products, and has effectively maintained its existing products. In response to its own interpretation of the marketplace, Cisco organized itself into six business groups: the core router group, the access router group, the workgroup products group, the ATM group, the IBM market group, and the Internet business group. That design worked well until the telecom meltdown occurred in 2000 and 2001. Faced with the same problems that other manufacturers faced —reduced order volume and massive inventories—Cisco regrouped and overhauled.

In August 2001, they announced yet another reorganization that converted their line of business structure to a centralized engineering and marketing organization. The result was the creation of 11 new functional organizations: access, aggregation, Cisco IOS router software, Internet switching and services, Ethernet access, network management services, core routing, optical, storage, voice, and wireless.

Additionally, Cisco has formed alliances with a number of *wide area network* (WAN) technology and service providers to develop flexible options for IP and FR services. Cisco has also expressed significant interest in the voice services marketplace.

Clearly, *Cisco the router company* could not in and of itself address the needs of such a diverse market. To date, the company has acquired the following corporations, along with the capabilities each provides:

- **Crescendo Communications**   *Fiber distributed data interface/copper distributed data* (FDDI/CDDI) hubs
- **Lightstream**   ATM switching equipment
- **Grand Junction Networks**   Ethernet and Fast Ethernet switches
- **TGV**   Intranet and Internet software
- **StrataCom**   ATM and FR switches
- **Nashoba**   Token Ring switches
- **Telesend**   D4 DSL frame multiplexers
- **Skystone systems**   SONET and *synchronous digital hierarchy* (SDH) equipment
- **Global Internet Software**   Security software for Windows NT systems
- **Ardent Communications**   Compressed voice, *local area networks* (LANs), and data over ATM and FR
- **Fibex**   Digital loop carrier equipment that handles circuit-switched and ATM voice traffic
- **Sentient**   Gateway equipment between ATM and circuit-switched voice
- **Amteva**   Unified messaging support software
- **GeoTel Communications**   Support for distributed voice call centers
- **Dagaz**   *X-Type digital dubscriber line* (XDSL) equipment

- **Light Speed International**   Voice-signaling support
- **Wheel Group Corporation**   End-to-end network security software
- **NetSpeed**   DSL equipment
- **Precept Software**   VoIP
- **CLASS Data Systems**   Priority resource control (QoS)
- **Summa Four**   VoIP switches
- **American Internet Corporation**   IP address management
- **Clarity Wireless**   Fixed wireless
- **Selsius**   IP PBX
- **PipeLines**   SONET/SDH routers
- **V-Bits**   Digital video over cable
- **WebLine Communications**   Internet e-mail routing software
- **Cerent**   ATM FR and IP over fiber
- **Monterey Networks**   Multigigabit optical router
- **Calista, Inc.**   Circuit and IP voice integration
- **IBM's Networking Hardware Division**   Switching and routing

Additionally, Cisco has enjoyed strategic alliances with a dizzying array of companies, including AT&T, Hewlett-Packard, IBM, GTE, Microsoft, Ameritech, and Sony. Recently, they announced a $1.05 billion investment in systems integrator KPMG Consulting.

If we analyze this remarkable collection of acquisitions and alliances, we discover that Cisco has become much more than a California-based router company. Ignoring the names of the companies for a moment, we find the following:

- ATM switches
- FR switches
- Ethernet, Fast Ethernet, and Token Ring switches

- DSL hardware
- SONET/SDH hardware and software support
- Network security
- Multiprotocol routing capabilities (including voice, video, LANs, and data)
- ATM-to-circuit-switched traffic gateway capabilities
- Telephony signaling (with IP to SS7 conversion)
- Unified messaging
- Call center support
- QoS support
- Wireless local loops
- IP and circuit-switched PBX capabilities
- VoIP switches

We are compelled to ask the following question: What is Cisco striving to become? Based on the previous list and the business units they have rallied around, they intend to be in the access and traffic aggregation business, the Internet-based switching and services business, the high-speed optical transport business, the metro network business, and a broad variety of other businesses. Obviously, Cisco intends be a full-service provider, and the company's alliances with integrators, standards bodies, content providers, local and long-distance telephone companies, consultancies, and other hardware providers reflect this fact.

COMPANY CONVERGENCE: AOL

The October 1999 issue of *Money Magazine* included an article entitled, "AOL: The One Stock You Can't Ignore." The October 29, 2001 issue of *The New Yorker* includes Ken Auletta's article, "Leviathan: How Much Bigger Can AOL-Time Warner Get?" Tumult has certainly characterized this corporation, but in the long run it has emerged as a shining success story among its peers, culminating with the acquisition of Time Warner in January 2000. The price, an almost unimaginable $154 billion, represents the largest merger in history, bigger even than the Sprint-MCI WorldCom megacorporation. The benefits that accrue to both corporations are staggering. The combined corporation will be in a position to offer a complete suite of services to customers, including both media and information content. AOL's 33 million customers who rely on AOL for web access will now also have access to Time Warner's vast content holdings. It also provides AOL with a broadband cable network of its own for distribution of AOL content.

Within the so-called dot-com world, AOL, barely 14 years old, is something of a silverback. It started in 1985 (the Internet equivalent of the late-sixteenth century) as Quantum Communications Services, which provided an interconnection service called Q-Link for users of Commodore 64 computers. Q-Link was followed in short order by AppleLink Personal Edition for Apple II users, PC-Link for Tandy Deskmate computers, and Promenade for IBM PS/2 users. In 1989, the name changed to America Online, and all services were discontinued so that the company could focus on their concept of a single product that would work for everyone. The initial service rolled out shortly thereafter for Apple II and Macintosh computers, and in 1993 a Windows version appeared.

"Buried in its success lie the seeds of its own destruction" is a phrase that characterizes AOL's rise to stardom rather nicely. In 1997, the company introduced a flat-rate pricing plan that, coupled with the enormous interest in the Internet and AOL's burgeoning customer base, contributed to severe congestion in telephone companies' central office switches and the inability of users to access the AOL network due to insufficient modems in the company's modem pools. After numerous court actions and nationwide appeals for relief from customers, the company beefed up its network and today enjoys revenues from 33 million customers worldwide.

AOL realized early on that the number of ISPs was growing, and that if they were to stay ahead of the competition curve they would have to offer more than simple access to the Internet. The company has a long history of crafting strategic alliances with hundreds of content providers who make their information available within the AOL domain so that subscribers do not have to leave AOL to reach it. Careful studies showed that a significant percentage of the company's subscribers never leave AOL—that is, they never actually use the button that enables them to go to the Internet itself because they find everything they need within the system. Consequently, AOL decided to make themselves over, becoming a full-service provider offering much more than simple Internet access. They became a portal provider, an e-commerce leader, and a provider of interactive chat capabilities. Through a series of carefully planned acquisitions and mergers, AOL was able to expand its footprint into a diversity of service areas. In addition to Time Warner, these acquisitions include the following companies:

- **Netscape**   Browser software
- **When.com**   Internet calendar and events software
- **PersonaLogic**   Interactive consumer buying guides
- **Mirabilis**   ICQ instant messaging and chat
- **CompuServe**   Sold the network, kept the customers!
- **Gateway ($800 million)**   Computer hardware
- **Digital Marketing Services**   Branded interactive services
- **MovieFone**   Movie listing and ticketing
- **MapQuest**   Online maps and travel information

The company has also formed strategic alliances with a number of companies, including

- SBC Communications
- Verizon
- Wal-Mart
- Circuit City

- VitaminShoppe
- eBay
- CBS
- Metrocall
- Packard Bell
- Telstra
- 3Com
- ABC
- *New York Times*
- E*Trade
- BBN
- Excite

AOL has clearly found a variety of ways to position itself before its existing and prospective customers, and take advantage of its strategic partners' abilities. By forming alliances with Verizon and SBC, AOL has provided good rationale to the ILECs for the deployment of DSL, which they have been slow to roll out in some cases. For $40 per month, a customer can have high-speed access to AOL services over the ILEC's DSL network.

Like Lucent and Cisco, the company exhibits a form of convergence best described as *heterogeneous*—that is, the company is engaging in mergers and acquisitions designed not only to expand the company's online service footprint, but also to move AOL into a variety of new, related services that complement one another in rather innovative ways. AOL's vast footprint as an ISP gives it and its content providers access to an enormous worldwide market, and their diverse services set attracts customers from all walks of life. A sampling of the services available from AOL includes access to nearly 100 major magazines; the Shopping Channel, which gives subscribers online access to more than 300 retailers; Starbucks, which provides coffee-related information; Thrive, which provides health-related information; NetGrocer, which allows subscribers to order groceries online; Digital City, which provides local news and information about a variety of cities; and, of course, the services that come packaged as part of the Time Warner deal. They have also made forays into long-distance telephone service and may soon offer network-based office automation applications, such as word processing, spreadsheets, and presentation software. As an ASP, they would no doubt capture a healthy proportion of the overall market—but *not* as a single-service ISP.

Of course, AOL has not been immune to the economic difficulties that have plagued the telecommunications services industry for the last six months. In August, AOL announced a corporate reorganization designed to integrate its online brands and web-resident properties as a way to capitalize on growth opportunities stemming from the growth of broadband demand. Notable elements of the plan include the creation of a new AOL interactive services group to include the traditional AOL online service, the company's Digital City enterprises, and a new vertical markets group charged with the development of programming, cross-brand advertising, and e-commerce opportunities in 10 key areas, including music, entertainment, personal finance, autos, travel, and a variety of others. The plan also includes strategies to consolidate the corporation's multiple web brands into a new AOL web properties group. These brands include Netscape, CompuServe, Moviefone, MapQuest, ICQ, and AOL instant messenger services. Another area that the plan will target is the creation of a new interactive marketing group that will oversee partnerships and sales revenues for all the AOL brands and properties to ensure that they are being leveraged as effectively as possible.

One very important aspect of the revised business plan is the intent to expand the role of the AOL broadband group and task the group with driving the consumer adoption of AOL services across cable, DSL, and satellite platforms. It is clear, therefore, that AOL understands where it is going and what it will take to get there.

THE SPECIAL CASE OF THE ILECS

The ILECs face competition from a variety of sources and must therefore take steps to protect their besieged marketshare and customer base. A number of technical factors have made them understand the steps they must take to remain competitive. These factors include

- Cable and wireless technology providers pose the greatest threats to the ILECs.

- Cable modems will dominate the market for high-speed Internet access, acquiring more than 50 percent of marketshare by 2005.

- Broadband wireless services such as *local multipoint distribution systems* (LMDSs) will become strong competitors in the next few years.

- As the ILECs lose marketshare, the cost to serve a shrinking customer base becomes prohibitive.

- Wireless and cable infrastructures are less expensive to implement than the traditional wire-based infrastructures deployed by the ILECs.

For the most part, the ILECs have exhibited a form of enhanced homogeneous company convergence, characterized by acquisitions of other local providers designed to expand their footprint because of the need for wide area data transport capabilities. SBC, for example, has acquired Pacific Bell, Nevada Bell, Ameritech, and SNET, while Verizon, the country's other powerhouse, has acquired NYNEX and GTE. These acquisitions are clearly homogeneous in nature; they add nothing new to the acquiring company's abilities other than geographical coverage.

Other activities, however, add to the competitive mix. Recognizing the strategic importance of wireless as an alternative access scheme, the ILECs' expansion into wireless properties has been spectacular. In September 1999, Bell Atlantic agreed to merge its wireless holdings with those of the United Kingdom's Vodafone Airtouch conglomerate, forming the largest mobile telephone company in the United States with 20 million subscribers and a combined net worth of $70 to $80 billion.

As part of the company's strategy to increase its wide area broadband presence, Bell Atlantic invested $2.2 billion in MetroMedia in October 1999. The investment will give the ILEC access to MetroMedia's fiber optic network in 50 major U.S. markets and a number of international locations as well.

SBC has been equally busy, forming alliances with AOL, DirecTV, and Prodigy. In November 1999, SBC announced plans to acquire a 42 percent equity stake in Prodigy Communications with plans to designate the service as its exclusive retail consumer and small business Internet access provider. As part of the overall deal, Prodigy agreed to use SBC as its preferred service provider for telecommunications and Internet services. In June 2000, Prodigy assumed management of SBC consumer and small business dial-ups, *integrated services digital networks* (ISDNs), and basic DSL Internet service customers. It also announced plans to relocate its corporate headquarters to Austin, Texas from White Plains, New York.

In January 2001, Prodigy and SBC renegotiated their agreement, extending their relationship through 2009. Under the renegotiated contract, Prodigy provides Internet content, news, and e-mail, while SBC provides network access, customer care, and distribution channels. As a further incentive, Prodigy receives $5 for each SBC DSL Internet customer.

Wireless access is another area where SBC has been successful. SBC's wireless service, offered through Southwestern Bell Wireless, Pacific Bell Wireless, SNET Wireless, Nevada Bell Wireless, Ameritech Cellular, and Cellular One, provides substantial coverage for local access on a close-to-nationwide basis. Ameritech, meanwhile, has formed an alliance with ITXC Corporation, an Internet telephony services provider that will now terminate calls on the Ameritech network.

USWest, recently acquired by Qwest, now forms part of a powerful combination of high-bandwidth, long-distance capability and local access providers in a number of major markets including Denver, Phoenix, Minneapolis-St. Paul, Seattle, Portland, Salt Lake City, and Albuquerque. The combined company can now provide long-distance voice and data transport, one of the major tenets of success in the evolving telecommunications market.

THE MISSING LINK

So, what are these companies missing, if anything? The ILECs offer narrowband and broadband wired and wireless local services, the ability to sell long distance as soon as FCC restrictions have been cleared, Internet access, and in some cases strategic relationships with content providers. What they do not have for the most part is a relationship with a systems integrator or an applications consultant. In the same spirit that caused Lucent to acquire INS, Cisco to acquire a piece of KPMG, and Novell to take a stake in IT consulting firm Whittman-Hart, the ILECs must round out their holdings by adding integration and consulting services to their physical network holdings. Being a large provider of access and transport, even as a sole provider, is not enough, particularly when faced with the fact that IECs and other market segments are aggressively looking to capture a substantial piece of the local access market. The critical importance of the *services* component of the telecommunications business is evident here, and knowledge-driven companies will win the game. If the ILECs are to protect their marketshare from the IEC intruders, they must offer a set of services that goes well beyond those offered by the competitors.

To do this, they must recognize and respond to three factors:

· *The ILECs must design a technology-based information and knowledge derivation infrastructure.* Although the secret to success lies in the ability to quickly and efficiently provision services and solutions, communications technology provides the critical underpinnings that make the services and solutions possible. The ILECs must therefore ensure that their networks are based on the right technologies, and that they are available throughout their operating regions. They must also build a knowledge management infrastructure and inculcate in all employees the importance of knowledge sharing. According to a report by the Conference Board[2] in New York City, "Getting people to share information may well turn out to be one of the key managerial issues of the next few decades."

- *The ILECs must be able to move quickly, to be flexible, and to adapt to change readily.* The ILECs must adopt managerial changes that will allow them to make decisions quickly and efficiently. This undoubtedly means evolving to more of a peer-to-peer model than they currently exhibit and moving decision-making power as low as possible within the organization. It is also a further reflection of the importance of knowledge-driven management. With the proper management of customer relationships and appropriate data-mining techniques, ILECs can anticipate marketplace changes and be prepared before the customer asks for them.

- *The ILECs must learn to sell solutions, not technologies.* Telephone companies have been driven by technology for so long that it is difficult to change the focus from technology-based selling to solution-based selling. It is an absolute necessity, however, if they are to remain in the game. As we have observed before, customers no longer care about the underlying technology. They do, however, care a great deal about whether the solution offered by the service provider will help them solve their business challenges and better position them in their own competitive markets, which means that the focus must be on the customer's business and application requirements.

---

[2]New York City Conference Board. "Knowledge Hoarding Is an Old Business Habit That Is Tough to Break." *Wall Street Journal*, 14 October 1999, page 1.

THE VIRTUAL CORPORATION: SUPPORTING TECHNOLOGIES

In order for the virtual corporation model to work properly, technologies must be utilized that create the illusion that the customer is doing business with a single company. Technologies must also help the various enterprise components of the virtual corporation work as a single entity.

When corporations join forces, an enormous amount of infrastructure work must be done if the merger is to function. Not only are there organizational issues to manage, but there are also daunting technical challenges to overcome. For example, corporate computer systems must be integrated, network infrastructures must be harmonized, applications must be joined, differing IT practices must be reconciled, and intracorporate communications must be assured. Furthermore, these exercises must be done as cost-effectively as possible, considering the close scrutiny that most mergers attract.

ENABLING THE VIRTUAL PRIVATE NETWORK (VPN)

Because the virtual corporation is comprised of distinct and often far-flung entities, and because it is common in modern corporations to support remote workers, technologies have evolved to support the critical strategic demand for the interconnection of multiple offices at a reasonable cost and for remote access to corporate information. The extremely competitive nature of the marketplace today requires that employees have immediate and efficient access to the most current information available, and that they be able to work remotely as efficiently as if they were directly connected to the corporate network. The other advantage has to do with extending the cloud. By creating a capable corporate network, customers can be brought in electronically to ensure the timely exchange of information, thus enhancing CRM activity.

One answer to this challenge is the *virtual private network* (VPN). A VPN is a cost-effective alternative to a dedicated facility. Instead of paying the mileage charges associated with private line, a VPN replaces the dedicated circuit with the public Internet, the use of which has no distance component associated with it.

VPNs represent an alternative to traditional dial-up access. In a direct-dial scenario, a remote worker dials directly into a corporate modem pool at headquarters, which means that the cost of the connection is dependent upon the distance the worker is from the headquarters facility. In most cases, the employee uses a toll-free number, but a cost is associated with this technique nonetheless.

VPN access is quite different. When remote users access a corporate network using a VPN, they do not dial directly into the corporate network. Instead, they create an Internet connection by dialing into a local ISP; then they use special protocols that create a tunnel through the fabric of the Internet that safely transports the information between the remote worker and the corporate network. Because the Internet is serving as the WAN, no distance-related cost is associated with this technique. The *free* Internet simply becomes an extension of the corporate network, providing access for remote workers. To protect the user and the company, secure protocols enable information to be passed safely between a remote employee and corporate computer systems. The difference between these two techniques is shown in Figure 2-5.

Design considerations must be taken into account when implementing a VPN versus a direct-dial solution. The first thing that must be determined is whether a VPN solution is merited. If the bulk of the users who will be dialing into the corporate network are geographically close to headquarters, a dial solution may be perfectly acceptable because they will incur no distance-related charges as a result of their sessions. If, however, the users are more geographically dispersed, then an Internet or alternative IP solution may be in order to keep the access costs down. The IT staff must perform an audit of all users to create a profile that will indicate to them whether a VPN solution is required.

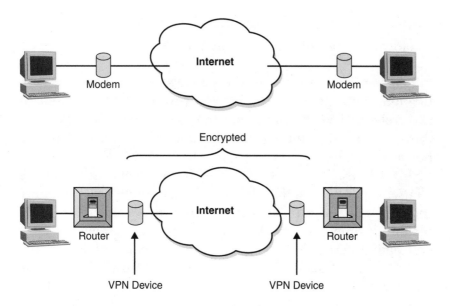

FIGURE 2-5   Direct-dial (top) versus VPN connectivity (bottom).

The second question that must be asked has to do with the number of ports that will be required on the headquarters server. For dial users, the IT staff must determine the number of modems that will be required, while VPN implementations are based on the number of simultaneous sessions that can be expected. Once the session count has been estimated, the total bandwidth required can be estimated based on the bandwidth required per session and the maximum number of sessions expected at any one time.

The third concern that must be addressed is network security. According to Gary Kessler, a specialist in network security, six questions must be answered when assessing security solutions:

· What are you trying to protect?
· Who are you trying to protect it from?
· What is the likelihood of an attack?
· What kind of attack is most likely?

- What are the results of an attack?
- How much protection can you afford?

The answer to these questions comes in three parts: authentication, data integrity, and session privacy. Authentication is the process of determining that the user really is who they say they are, and that the data being tunneled into the corporate network is from a reputable source. Because VPNs use the public Internet as their transport fabric, it is critical to authenticate all users and the data that they send to corporate systems before they are allowed in. This can be done in a number of ways, including simple login/password combinations, secure tokens, and digital certificates. Login/password combinations are the easiest form of security to break because most users choose passwords that they can easily remember—which means that a hacker can more easily determine what they are. One young hacker that I interviewed for an article on network security proudly boasted that "given five minutes with your wallet or purse, I'll have enough information to guess your passwords."

A more robust alternative to passwords is the secure token. Secure tokens are based on one of two factors: *something you have*, such as a secure ID card or a one-time password, or *something you are*, such as a fingerprint scan, retina scan, or voiceprint. These techniques are becoming more common, but still fall on the expensive side of the security spectrum.

Although passwords and secure tokens offer a modicum of security, digital certificates have emerged as the clear winner. The concept of a digital certificate is quite simple. It is nothing more than a digital document that contains a public key used to read secure documents, a name associated with that key, an expiration date, the name of the trusted third party that issued the certificate, a serial number, and information about appropriate use policies for the certificate. It may also contain the user's digital signature.

The certificate is used for the following purposes: to establish the identity of the user by associating a public key to an individual or organization, to assign authority by clearly stating the actions that the holder of the certificate may or may not take, and to protect the data that will be transmitted by encrypting the information using one of several secure protocols.

VPNs, then, represent a secure and cost-effective alternative to a dedicated network, making it possible to integrate far-flung corporate LANs and WANs using a public network infrastructure. Currently, private line and FR represent the bulk of all WAN installations. Unfortunately, they are relatively inflexible, a problem in today's dynamic network environment. Because they are usually sold under a three-to-five-year contract, they have become less attractive for customers who do not want to be locked into a network solution that is not capable of evolving in lockstep with their changing applications and geographic footprint. Thus evolved the strong interest in VPNs.

The cost advantages to VPNs are quite significant. In addition to the savings that result from the elimination of private-line distance charges and dial access costs, savings also occur due to reductions in capital equipment and support activities. Because VPNs use a single WAN interface for multiple functions, the data that would normally have passed through several data devices now requires one. Support costs are reduced because of the ability to consolidate all support functions within a single help-desk organization.

Of course, there are challenges associated with VPNs, not the least of which is performance. The Internet is not known for providing rock-solid QoS, and given that it constitutes the heart of the VPN, a number of questions arise. An ideal IP-based VPN would

- Be ubiquitously available, secure, and reliable.
- Offer a variety of network management and billing options.
- Provide measurable SLAs.
- Enable the user to differentiate QoS on a per-flow and/or per-application basis.

The Internet, or even a pure IP network, does not inherently have the capability to do all these things. IP on top of an ATM backbone, however, does have the capability, and this model seems to be the emerging choice for carriers looking to deploy QoS-capable public networks. By associating IP addresses to ATM virtual circuits, QoS can be assured.

So, what is the future of the VPN? First, it is important to acknowledge that the private-line network will never disappear, but as Internet QoS performance and encryption protocols improve, the need for a dedicated facility will become less and less obvious. The VPN, because of its capability to reduce costs, support new business opportunities, and improve flexibility and speed to market will be recognized as a powerful enabler of CRM and therefore of competitive positioning.

THE EVOLVING COMPUTER

The evolution of the modern computer from the centralized mainframe to the fully distributed system has occurred in concert with the decentralization of the modern corporation and its growing focus on the need to reduce the distance between the company and the customer.

Contrary to what many believe, the mainframe computer is far from dead. Although it is no longer the center of the computing universe, it still plays a major role for corporations with the need for centralized database and application support across the enterprise. Neither minicomputers nor networked PCs can provide this capability, so although its role has shifted, the mainframe still plays a central role in enterprise computing.

When IBM first rolled out its corporate mainframes several decades ago, they relied on a hierarchical approach to computing that emulated the managerial model of the corporations that installed them. *Systems network architecture* (SNA), released in 1974, followed the same model in which all roads led to the mainframe. Today SNA is thriving, but it is a very different beast than the centralized approach that first rolled out of IBM's doorway. Recognizing the evolution to a distributed computing model, IBM modified SNA to accommodate the needs of distributed corporations. *Advanced program-to-program communication* (APPC) enables communication between peer-level programs without having to create a session through the mainframe, a first step away from the hierarchy of yore. *Advanced peer-to-peer networking* (APPN) came shortly thereafter, enabling SNA to support the requirements of distributed computing. With APPN, minicomputers can be networked together to create a true peer-to-peer environment, moving IBM from its centralized roots into the departmental computing model while maintaining its position as the leading provider of mainframe technology.

A concern generating significant angst within IT departments these days is the issue of legacy integration. Although many corporations have significant investments in mainframe systems, there is tremendous pressure to integrate newer technologies, including IP. This represents another example of the adage if it ain't broke, don't fix it. The fact that a corporation has legacy systems in place does not necessarily imply that they are outmoded. Many corporations have routinely maintained and upgraded their systems such that they are still high-performance machines. In fact, according to the META Group consultancy, most companies have found that it makes more sense to maintain legacy installations than to replace them with modern technology because 70 percent of all corporate data still resides on them.

However, pressure from the *Transmission Control Protocol/ Internet Protocol* (TCP/IP) is forcing IT managers to seriously consider merging it into their SNA-based systems. There are good reasons for the inclusion of TCP/IP, including its ubiquity, its use in thin client systems, its flexibility, its error-recovery capabilities, and its enormous installed base, thanks to the popularity of the Internet. The greatest benefit, of course, is the universal nature of TCP/IP as a standard network protocol, and the fact that it is now included in every network operating system is an indication of its central role in modern networking.

TCP/IP was first introduced into the AS400 operating system, which allowed it to be easily integrated into the corporate computing environment. The challenge that had to be overcome in the newly integrated network was the need to route SNA traffic over a TCP/IP WAN network because of the continued presence of legacy applications. A number of solutions have been proposed, most of which involve tunneling or encapsulation. These techniques involve the encapsulation of SNA data and include *data link switching* (DLSw), which is the most common technique used. In DLSw, routers perform the actual encapsulation process, meaning that the end devices only need to handle SNA traffic with no need to understand TCP/IP. As a result, the integration of legacy environments with TCP/IP is relatively straightforward and easily accomplished by competent IT staff.

Of course, the computing world has continued to evolve beyond the mainframe. The PC's impact has been immeasurably profound and, combined with high-speed LAN technology, enables the creation of massively distributed virtual supercomputers throughout a corporation.

STORAGE AREA NETWORKS (SANS)

One of the principal advantages of a mainframe is the ability to manage and maintain a massive centralized database, accessible by all users. One recent development is the SAN. According to the research firm *International Data Corporation* (IDC), storage area networks and the Fibre Channel[3] technology that interconnects them are fast becoming the preferred storage technology for enterprise data centers. In fact, according to work performed by the Enterprise Storage Group, the storage industry is growing by 65 percent annually, and storage applications are contributing significantly to the overall volume of network traffic.

The problems that SANs are designed to correct are common. In distributed environments, data are stored on client systems, managed servers, departmental servers, and mainframes. This results in a number of problems, including the difficulty to

---

[3]Fibre Channel is an *American National Standards Institute* (ANSI) standard that defines a high-speed technique for transmitting data between all manner of data devices. It transmits at speeds ranging from 133 Mbps to 1 Gbps, and 4 Gbps systems are under development. In spite of the name, Fibre Channel can operate over either fiber or coaxial cable. See **www.fibrechannel.com** for more information.

universally access corporate information, uncoordinated database backups, a lack of version control, the inability to manage and monitor performance, and uncontrolled storage costs. In a SAN, data storage is centrally managed to guarantee fault tolerance, scheduled backups of critical data, flexibility, and universal accessibility.

Although SANs offer a great deal to IT managers, problems are still being worked out of this relatively new technology. First, there are interoperability issues to be resolved on both the client and server sides of the equation. Although major vendors have announced Fibre Channel functionality, some third parties do not offer compatibility with the technology. Another limitation is distance. Currently, Fibre Channel is limited to approximately six miles, although a 25- to 30-mile enhancement is in the works. As Figure 2-6 illustrates, users attached to LANs or other components of the corporate network can access any of a number of database components via the Fibre Channel hub, which provides universal, high-speed connectivity.

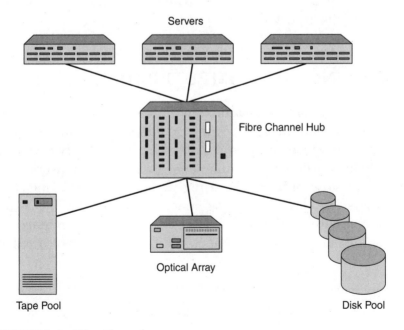

FIGURE 2-6   Fibre Channel.

A number of major vendors have announced SAN products, including Hewlett-Packard and IBM. Furthermore, a new type of company, called a *storage service provider* (SSP), has emerged. Originally offering storage rental, these companies now manage storage devices for enterprise customers.

THE NEED FOR MOBILITY

Equally powerful has been the response of the computer industry to the need for mobility. The laptop computer is so common today that it is no longer a novelty, but the evolution doesn't stop there. The dramatic impact of palm-top computers, most notably the 3Com Palm Pilot, has created a whole new realm of computing capabilities. Today, knowledge workers and road warriors often travel with nothing more than a palm-top and a cell phone with which they can accomplish most of the same tasks that formerly required an office, computer, and phone. Collaborative vendors have already released devices that combine the capabilities of a palm-top and a cell phone, and further innovation is expected.

THINNING THE COMPUTER HERD

Another ongoing trend—although there is some confusion over whether *ongoing* means growing or shrinking—is the development of the so-called network computer, led by Larry Ellison of Oracle fame who first attempted to succeed with network computers in the mid-1990s using Oracle software and Sun Microsystems' Java[4] operating system. The idea behind a network computer is that the bulk of the intelligence required to make the machine useful resides in network servers rather than

---

[4]Java is a programming language invented by Sun Microsystems in 1995. It is designed to support distributed computing environments where applications reside on centralized web servers that can be downloaded across the Internet to a PC where they execute locally, thus eliminating the unpredictable latency of the Internet in the middle. Java is therefore ideal for distributed computing.

in the PC itself, thus making the client device inexpensive. From the point of view of the corporate IT manager, network computers (sometimes called *thin clients*, another name from Oracle) represent the possibility of substantial savings because the PCs that are distributed to the corporation's many users are slimmed-down versions of a PC and are thus far less expensive. The intelligence resides in the servers that are shared among all users, making administration and maintenance, and therefore cost control, significantly simpler.

SUMMARY

Driven by the need to provide a suite of services through consolidated technologies, company convergence has raged throughout the telecommunications industry. As companies struggle to become the chosen single-source provider for telecommunications services, the deal-making accelerates as they form competitive clusters that can meet the challenges posed by customers in a marketplace that grows more competitive with every passing day. However, the clustering of companies is only a piece of the convergence process. They must also take whatever steps they can to ensure that they are functionally converged as much as they are organizationally converged. Otherwise, the customer will not benefit.

This section focused on the evolution of the electronic business, an inevitable evolution in the Internet-driven business model. There are, however, pitfalls that corporations must be aware of as they progress with their plans to become fully-functional e-businesses. According to the Gartner Group, 75 percent of all e-business conversion projects will fail due to poor business planning and a lack of understanding of the technology required to bring about the evolution. The five most common reasons for failure include poor project management, a lack of clearly defined goals, inflexibility, minimal knowledge of competitors with regard to their e-business activities, and a sense that the conversion to e-business is an end unto itself, rather than a means to an end.

The traditional business model is changing, replacing wholesalers and retailers with online intermediaries and web merchants. Those intermediaries are becoming specialized, focusing on specific industries or business processes that they have the ability to positively affect. Their services include service hosting, transaction mediation, the aggregation of buyers and sellers, and the provisioning of "safe harbors" for buyers and sellers. They serve as hubs, and their value is based on Metcalfe's Law, described earlier in the chapter. Clearly, the more intermediary hubs that exist, the greater the aggregate value of the network and its services. These hub services generate significant revenue.

The main challenge that these new service providers will face is the need to diversify into specific market segments. They have a choice: They can diversify vertically, which simply means that they become industry-specific, or they can diversify horizontally, which implies that they focus on business processes and work to make them more efficient, regardless of the industry. If a company chooses to diversify vertically, their success will increase as a function of a number of key factors. These include the fragmentation between buyers and sellers, the ability to provide a sophisticated online search capability, discrete knowledge of the industry they serve as well as the relationships that exist between the various players, supply-chain efficiencies, and the ability to aggregate major suppliers with buyers. Similarly, horizontal diversification succeeds as a function of process standardization, a discrete understanding of the business process of the industry served, knowledge of process automation, and the ability to customize services and products to serve specific industry niches.

In the next section, we look at the final component of the convergence phenomenon—the convergence of services. There we reach the climax of the convergence process and pull it all together.

PART THREE

SERVICES CONVERGENCE

Recognize that every out-front maneuver you make is going to be lonely and a little bit frightening. If you feel entirely comfortable, then you're not far enough ahead to do any good. That warm sense of everything going well is usually the body temperature at the center of the herd.

—Anonymous

Competencies owned and nurtured by a company represent its critical resource and competitive advantage, and the company should create a portfolio of services that contribute to and extract value from those competencies.

—G. Hamel and C. Prahalad

Putt's law: Technology is dominated by two types of people: those who understand what they do not manage, and those who manage what they do not understand.

SERVICES CONVERGENCE

In previous sections, we described the overall convergence process. Initially, the collection and analysis of customer interaction activity lead to an understanding of their current and future service requirements. Those requirements translate into an understanding of the technology necessary to satisfy those service requirements, but because service providers do not typically have all the needed technology and know-how, they acquire it through mergers, acquisitions, or strategic corporate alliances. Through those activities, they form virtual corporations that, if built correctly, have the ability to not only deliver, but to anticipate future customer service requirements. In a market as competitive as the telecommunications industry, the ability to differentiate products that are, for all intents and purposes, commodities becomes a matter of survival.

In this part, we delve into the end game—the actual services that customers are looking for and the technologies required to deliver them. As before, we look to the future and list the key forces that affect the convergence of services in the new corporation. Technology convergence is a facilitator, while company convergence is more of a convenience mechanism.

We now turn our attention to the third aspect of the convergence troika, services convergence. Here we must observe one fundamental truth: The seat of telecomm power is shifting away from the service provider and moving closer to the customer. Success in this marketplace lies with the ability to deliver a complete solution that satisfies customer requirements for competitive positioning. This is the promise of services convergence, and it is supported by the very capable pillars of technology and company convergence.

Services convergence means that a company must offer much more than pure technology if it is to be successful in the long term. The key to success isn't the technology—it's what is *done* with the technology that really matters. Content, applications, and access appliances are the drivers; technology is the facilitator of those drivers. In an ideal situation, the technology itself is invisible, while digital media, broadcast content, online publishing, and digital copyright protection become *very* visible —and lucrative. Services, solutions, and competitive advantages become the *real* goals for businesses in the digital economy. And the most important differentiator of all is *quality of service* (QoS), the measure of the degree to which offered applications and services transported across a network infrastructure satisfy customer requirements. It is *not* measured in terms of network availability, response time, or mean time between failures because those are measures that service providers care about. QoS must be measured in terms defined by the customer.

STORAGE AREA NETWORKS (SANS) WILL BECOME INCREASINGLY IMPORTANT

In concert with the core-to-edge migration of intelligence and bandwidth, *storage area networks* (SANs) will take on an increasingly important role as the network's role itself changes. Furthermore, the services set that can be offered by an edge-intelligent network is quite broad and provides a significant enabler for bandwidth companies wanting to broaden their offered services set. Working in harmony with high-speed, low-cost transport technologies such as InfiniBand, SANs will become powerful facilitators of service evolution, especially among such segments as the *application service provider* (ASP) market.

THE SERVICES

I've looked at clouds from both sides now,
From up and down, and still somehow
It's cloud illusions I recall.
I really don't know clouds at all.

Joni Mitchell, "Both Sides Now"

It is important to remember that the service provider's view of services differs substantially from that of the customer. To be innovative and competitive, the service provider must view the services' world through the eyes of its customers. Otherwise, it becomes far too easy to think of service and technology as synonyms for each other, which they most definitely are not. As we have said before, technology is a facilitator of service and, as Joni Mitchell's song suggests, should be completely invisible to the customer. All they should see is an opaque cloud, from which are delivered the capabilities they require to be competitive within their own markets.

From the customer's perspective, services include

- Voice
- Distributive video (television or *video-on-demand* [VOD])
- Interactive video
- Image transport
- Narrowband and broadband data services
- Internet access
- E-mail
- Fax

Ideally, a service provider delivers them over a single or limited number of network interfaces, guarantees that the quality of the delivered service will be in accordance with a *service level agreement* (SLA) that both parties have signed, and is always one step ahead of the customer in terms of providing for their telecommunications needs.

From the point of view of the service provider, several key components must be addressed if they are to meet their customers' demands, the most important of which is QoS. To a customer, QoS means many things, including

- Intelligible, recognizable voice
- A wide selection of VOD movies

- High-quality videoconferencing
- Medical image-quality transport
- A variety of data services, ranging from low to high bandwidth
- Anywhere, anytime Internet access
- E-mail
- Low-cost fax (possibly IP-based)

To a service provider, the concept of service takes on different meanings because they must concern themselves with the technology that will be required to provide all the capabilities described above. Their key concern is to build a network infrastructure that has the capacity, flexibility, service granularity, *growability*, and robustness to meet customer requirements. In their terms, QoS may mean

- Adequate, easy-to-provision bandwidth
- Minimal delay
- Minimal jitter
- Minimal information loss
- Scalability of network resources
- Simple interworking with other networks
- Enhanced management capabilities
- Security
- Reliability

As we discussed in an earlier part of the book, there is significant interest in migrating to an IP network infrastructure because of the universality and flexibility that it enables. However, we also observed that although IP has significant advantages for service providers and customers alike, there are some downsides associated with it, not the least of which is its inability to provide QoS. Consequently, we now see service providers building multipurpose *asynchronous transfer mode*

(ATM) core networks, over which they run IP at the network layer.

This hybrid approach has a number of advantages. First, it takes advantage of ATM's capability to provide granular QoS, seamless integration with *Synchronous Optical Network/ Synchronous Digital Hierarchy* (SONET/SDH), and well-defined interfaces with a variety of access technologies. Its widespread international acceptance is also an advantage for global corporations.

Second, this approach enables IP's many advantages to be integrated into the mix. IP offers a universally accepted addressing scheme and the capability to interoperate with a broad range of widely deployed protocols. That, combined with the bandwidth and QoS of ATM, makes for a powerful and enormously flexible technology combination. If a service provider were to deploy ATM and IP as their core multiservice network, they would be able to offer access and transport services for any imaginable application over a single network infrastructure, thus realizing their own requirements for achieving higher service quality at a lower cost.

VIEWING THE NETWORK: INSIDE OUT OR OUTSIDE IN?

There are two ways to view the network: from the perspective of the service provider or from the point of view of the customer. The service provider sees a group of access technologies that will connect to their core network, while the customer sees a transport cloud that his or her access technologies will derive services from. The network, then, is comprised of two distinct areas of responsibility: the *edge* and the *core*. The edge of the network is responsible for aggregating traffic, prioritizing it, and packaging it for handoffs either from the customer to the cloud or from the cloud to the customer. The core is responsible for maintaining the QoS requested by the customer's application on an end-to-end basis by managing the nine QoS factors mentioned earlier: bandwidth, delay, jitter, information loss, scalability, interworking, network management, security, and reliability. We will discuss each of these in turn.

Bandwidth is exactly that—a measure of the volume of information that can be transmitted through a channel in a given period of time, usually a second. *Delay* is a measure of the amount of time it takes for a packet to be transmitted over a network. As long as the delay is constant, that is, all packets are delayed the same, problems are minimized. When the end-to-end delay begins to be undependable, however, problems become significantly worse. This variability in transit time is called *jitter*, which measures the variability in delay from packet to packet between two endpoints.

Information loss is an indirect measure of the quality of the overall network. If packets are being lost on a regular basis, service quality suffers and delay may increase due to the need to resend the missing packets. *Scalability* is an indication of the ease with which the network accommodates growth. *Interworking* reflects the degree to which the network is designed around international standards and is therefore interoperable with other networks.

Network management is critical from the point of view of the service provider because without network management the network cannot be properly monitored to determine whether it is performing up to the requirements of the SLA.

Security is an important selling point in any network, particularly when *virtual private networks* (VPNs) are to be implemented across a public network infrastructure. Finally, *reliability* is the principal factor when negotiating SLAs. To the customer's way of thinking, all the previous factors fall into this category; from the service provider's point of view, however, reliability is a measure of the network's technical performance.

The bottom line, then, is that customers have certain preconceived expectations about what they should receive from their service provider. As long as they understand those expectations, the service provider will be able to anticipate those requirements and build a network infrastructure that can handle them. Today, IP is emerging as a major component of the so-called next-generation network for many good reasons that include economics, diversity, ubiquity, scalability, and capability. IP telephony is a rallying cry for IP's widespread deployment; in the next section, we examine how it works, how it differs from traditional telephony infrastructures, and how it will be implemented as a viable contender for voice service transport.

The traditional legacy telecommunications network consists of two main regions that can be uniquely and clearly identified: the network itself, which provides switching, signaling, and transport for traffic generated by customer applications, and the access loop, which provides the connectivity between the customer's applications and the network. In this model, the network is considered to be a relatively intelligent medium, while the customer equipment is usually considered to be relatively "stupid."

Not only is the intelligence considered to be concentrated in the network, but so too is the bulk of the bandwidth because traditional customer applications don't require much of it. Between switches, and between offices, however, enormous bandwidth is required.

Today, this model is changing quickly. Customer equipment has become intelligent, such that many of the functions previously done within the network cloud are now done at the edge. *Private branch exchanges* (PBXs), computers, and other devices are now capable of making discriminatory decisions about required service levels, obviating the dependence upon the massive intelligence embedded in the core.

At the same time, the bandwidth is moving from the core toward the customer, as applications evolve to require it. There is still massive core bandwidth within the cloud, but the margins of the cloud are expanding toward the customer.

The result of this evolution is a redefinition of the regions of the network. Instead of a low-speed, low-intelligence access segment and a high-speed, highly intelligent core, the intelligence has migrated outward to the margins of the network, and the bandwidth, once exclusively a core resource, is now equally distributed at the edge as well. Thus, we see something of a core and edge region developing in response to changing customer requirements.

THE EVOLUTION OF IP TELEPHONY

IP telephony promises to reduce communication costs and simplify multiservice corporate networks because of its packet-based underpinnings. However, to fully understand IP telephony and all that it and other IP-based services have to offer to the service provider as well as to the customer, it is first important to understand the existing telecommunications network that it may someday replace.

When a call is placed via a traditional telephone network, the process of call connection is simple and straightforward. The caller notifies the local switch of his or her intent to place a call by lifting the handset, which completes a circuit and allows a current to flow from the switch to the telephone.[1] The network responds to the caller's request for service by sending a dial tone from the switch to the telephone. The caller then notifies the switch of the destination address by entering the number to be dialed; the switch responds that it has received the appropriate number of digits by putting a ringing tone, a busy signal, or a fast busy signal in the caller's ear, as appropriate.

Of course, there is much more to it than that going on behind the scenes. When the switch receives the dialed digits, it must first determine whether the call is local or long distance, which it does by consulting a table that identifies local NXXs (prefixes). In the number 508-555-1798, 555 is the NXX. If the call is local and served by the same switch, the switch connects the call directly. If the called party is served by another switch, whether local or long distance, the originating switch must reserve a trunk to connect the originating and destination switch, and then send the call to the destination switch. If the destination switch is in a different geographic area, the call must first be handed off to a long-distance company, which will in turn hand the call back to the receiver's local service provider upon call completion.

---

[1] We will assume that this is an analog local loop and ignore *Integrated Services Digital Network* (ISDN) for the moment. The concept, however, is the same; instead of a dial tone and other analog constructs, ISDN uses a variety of packets to control the call setup process.

The typical telephone network is comprised of a hierarchy of local and toll switches that provide local interconnections between customers and the service provider, and ultimately between the customer and their chosen *interexchange carrier's* (IEC) network. The switches, in turn, are interconnected with an interstitial fabric of optical multiplexers, digital cross-connect devices, SONET or SDH add-drop multiplexing equipment, and a variety of other components. The entire system is controlled by a separate signaling network known as *Signaling System 7* (SS7), which is the protocol suite responsible for setting up, maintaining, and tearing down a call. SS7 also provides access to enhanced services such as *Custom Local Area Signaling Service* (CLASS), 800-number portability, *local number portability* (LNP), *line information database* (LIDB) lookups for credit card verification, and other enhanced features.

The original concept behind SS7 was to separate the actual calls on the *Public Switched Telephone Network* (PSTN) from the process of setting up and tearing down those calls, as a way to make the network more efficient. This had the effect of moving the intelligence out of the PSTN and into a separate network where it could be somewhat centralized and therefore made available to a much broader population. The SS7 network, shown in Figure 3-1, consists of packet switches (*signal transfer points*, or STPs) and intelligent database engines (*service control points*, or SCPs) interconnected to each other and to the actual telephone company switches (*service switching points*, or SSPs) via high-speed digital links, typically operating at 56 to 64 Kbps.

When a customer in an SS7 environment places a call, the following process takes place: The local switching infrastructure issues a software interrupt via the SSP so that the called and calling party information can be handed off to the SS7 network, specifically an STP. The STP in turn routes the information to an associated SCP, which performs a database lookup to determine whether any special call-handling instructions apply. For example, if the calling party has chosen to block the delivery of caller ID information, the SCP query will return that fact.

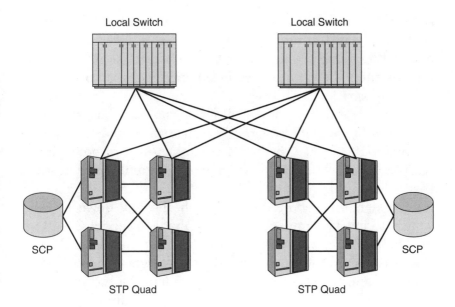

FIGURE 3-1   The SS7 network.

Once the SCP has performed its task, the call information is returned to the STP packet switch, which consults routing tables and then routes the call. Upon receipt of the call, the destination switch will determine whether the called party's phone is available, and if it is, it will ring the phone. If the customer's number is not available due to a busy condition or some other event, a packet will be returned to the source indicating the fact, and SS7 will instruct the originating SSP to put a busy tone or reorder in the caller's ear.

At this point, the calling party has several options, one of which is to invoke one of the many CLASS services such as *automatic ringback*. With automatic ringback, the network will monitor the called number for a period of time, waiting for the line to become available. As soon as it is, the call will be cut through, and the calling party will be notified of the incoming call via some kind of distinctive ringing.

Thus, when a call is placed to a distant switch, the calling information is passed to SS7, which uses the caller's number, the called number, and SCP database information to choose a route for the call. It then determines whether there are any special call-handling requirements to be invoked, such as CLASS and instructs the various switches along the way to process the call as appropriate.

These features comprise a set of services known as the *Advanced Intelligent Network* (AIN), a term coined by Telcordia (formerly Bellcore). The SSPs (switches) are responsible for basic calling, while the SCPs manage the enhanced services that ride atop the calls. The SS7 network, then, is responsible for the signaling required to establish and tear down calls, as well as to invoke supplementary or enhanced services.

One problem with Telcordia's concept of the AIN is the fact that it is somewhat incomplete. It was designed to be an open standard and therefore vendor independent. Unfortunately, by the time AIN arrived, there was already a substantial embedded base of legacy equipment, which meant that to implement AIN, equipment manufacturers had to modify their in-place hardware. This resulted in a significant expense and posed a serious obstacle to service providers wanting to deploy new services. In addition to that, the network management capability of the AIN was incomplete. The ability for service providers to manage existing services or to create new ones was difficult, and this factor slowed deployment as well. For these reasons, IP telephony has caught the attention of service providers who must move quickly if they are to be competitive in a marketplace characterized by new, low-cost, and aggressive entrants who do not have the burden of a legacy architecture to deal with.

In the IP environment, the goal is identical, but a different process is used. SS7 will still be used in the IP world, but the SS7 packets will be converted to IP-based signaling packets by some form of intelligent peripheral. A number of efforts are underway to directly interface IP to SS7, including Nortel Networks' IPS7 and similar protocols from other vendors.

With that understanding in mind, let us now consider how a long-distance IP telephone call might be processed, using an *Internet telephony service provider* (ITSP). In Internet telephony, we must introduce a number of new devices (see Figure 3-2) and eliminate some old ones.

Because Internet telephony will exist in parallel with traditional circuit-switched telephony for many years to come, local and long-distance switches will remain a reality in the circuit-switched domain. In IP telephony, however, we now introduce *gateways* and *gatekeepers*. A gateway is a device that performs destination address resolution, connection monitoring and management, circuit-to-packet conversion, voice digitization and compression, circuit-to-packet signaling, security, and

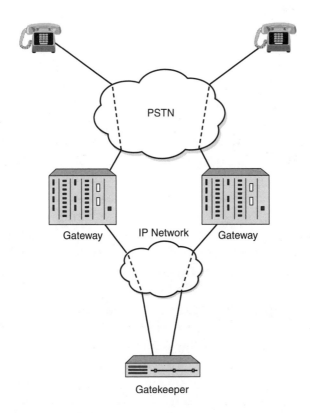

FIGURE 3-2   Interconnecting the PSTN and IP networks.

accounting. Some gateways are robust enough to replace a local switch; all of them replace the functions provided by a toll switch. A gatekeeper, on the other hand, serves as the administrative authority in IP voice environments, managing such functions as bandwidth management, call-zone management, address translation, and usage or admission control.

Call-zone management simply means that the network can be logically managed such that certain calls are restricted or otherwise controlled to satisfy the demands of management.[2] Address translation refers to the fact that a user may be able to place a call to another user by entering their e-mail address. The gatekeeper will translate the address into a telephony address that the network can handle. Finally, admission control is simply a security/management function that enables calls to be managed based on the time of day, user authentication, address pooling, or other methods.

The most desirable benefit of IP telephony is its capability to use the Internet or a higher-quality IP network as a means to bypass the traditionally expensive long-distance network. Thus, the call might proceed as shown in Figure 3-3. As with a traditional circuit-switched call, the caller goes off-hook and dials a long-distance number. When the local switch receives the dialed digits, it recognizes that the call is not local and proceeds to hand it off to the customer's chosen long-distance carrier. In this case, the customer's long-distance provider is an ITSP, so the circuit-switched call is handed off to the ITSP, which converts the stream of information to a stream of IP packets that it must now route to the requested destination, all the while being cognizant of the fact that this is a voice call and must therefore be afforded strictly controlled QoS.

The ITSP must first resolve the address so that it can route the call to the proper gateway that interfaces with the local service provider that serves the destination customer. Once it has determined the address, it must route the call, keeping in mind

---

[2]A number of protocols are often used to control these functions and to establish the zones within an IP telephony network. H.323 is common, although *Session Initialization Protocool* (SIP) and *Media Gateway Control Protocol* (MGCP) and their "relatives," discussed a bit later, are gaining fast in popularity.

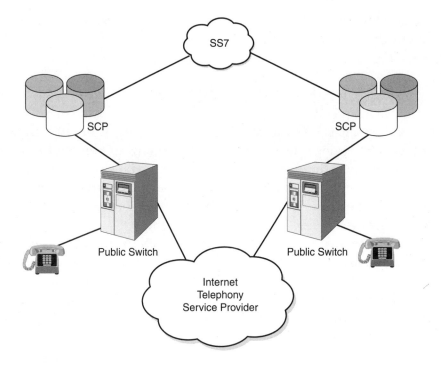

FIGURE 3-3   IP telephony.

that low delay is of paramount importance. At the same time, the ITSP must signal to the far end to ensure that the call can be connected through the called party's local service provider. The overall goal is identical; it is simply handled in different ways.

The greatest difference between circuit-switched voice and IP voice is the issue of quality control. In circuit switching, QoS is largely guaranteed because of the nature of the infrastructure. In IP voice, however, there are no guarantees, again because of the nature of the connectionless IP infrastructure. Thus, something must be done to ensure quality control if voice is to be transported over an IP infrastructure with anything resembling carrier-class service. One solution, of course, is to rely on a high-quality transport infrastructure such as ATM, which many companies are now doing. The alternative is to use emerging QoS protocols and devices capable of responding to their mandates. Those protocols and the devices that implement them are the subject of the next section.

EMERGING QOS PROTOCOLS

A conflict exists today in the area of QoS protocols because they tend to be either Internet driven or telephony driven. Not until recently have there been efforts to reconcile the differences between the two and to move toward a single common foundation of standards that serve the requirements of both the circuit-switched and packet-switched camps.

For a service provider to guarantee QoS for the purposes of SLAs, it must be able to manage both call signaling and content queuing. The signaling is required not only for the setup and teardown of calls, but also for the QoS requests that are an inherent part of signaling in a connectionless, IP-based network. Furthermore, every device along the flow of the data must be able to acknowledge and support the guarantee of delay and bandwidth. If a device is incapable of supporting the requested quality level, the call is refused.

To guarantee that delay will be minimized, switches and routers must be able to buffer flow state information—that is, the characteristics of each flow of information headed into or out of the network. Without that capability, the network cannot guarantee minimal delays, which means that delay-sensitive services such as voice and video—clearly two cash-cow services —cannot be provisioned, and SLAs cannot be met.

To guarantee bandwidth, switches and routers must support priority queuing and some assurance that the queues will not be oversubscribed. Unfortunately, many vendors offer priority queuing, but because IP has no way to prevent oversubscription, they have no mechanism to complete the picture. A number of protocols have emerged that address this concern and will be discussed later in this chapter.

Two major QoS mechanisms have emerged for controlling quality guarantees across the network. The first is called

Explicit QoS, which gives applications the capability to request specific levels of service quality. Routers or switches are then expected to make every effort possible to accommodate the request. The *Resource Reservation Protocol* (RSVP), discussed later, is a good example of an Explicit QoS technique.

The second approach is called Implicit QoS, in which routers and switches automatically allocate QoS resources, such as bandwidth based on criteria established by a network administrator. These criteria include such things as an application type, a logical port address, or a protocol ID. *Differential service* (DiffServ) and ATM, also discussed later, are good examples of Implicit QoS.

To date, ATM has assumed the role as the ultimate QoS provider across the wide area. We have also noted, however, that it is not necessarily the ultimate long-term choice because of such factors as its tendency to be expensive, overhead-heavy, and somewhat complex. The ultimate question in all of this is whether QoS is sustainable across a *wide area network* (WAN) using protocols other than ATM. The answer, of course, is yes, although there can be a significant cost associated with doing so. The cost is often paid in bandwidth because to achieve sustainable QoS, networks have to be either overengineered or undersubscribed because protocols for intelligently provisioning and guaranteeing QoS are not quite ready for prime time.

There is a great deal of discussion these days in the trade journals by technicians who claim that the glut of bandwidth that is upon us makes it feasible to overengineer without a sense of waste. And although this is true from a brute-force point of view, the philosophy is flawed. Parkinson's Law (work expands to fill the time available) is alive and well in telecommunications, and although there may be an overabundance of bandwidth today, that will not be the case in the future. Thus, throwing bandwidth at the problem as a solution is less than ideal, and far from elegant. ATM provides the elegance today; other protocols will inherit the crown over time.

Quality is affected by a number of factors including latency or overall delay, echo cancellation, call-completion ratios, and post-dial delay, to name a few. One way around the challenge of dealing with the unpredictability of the Internet is to use a general technique called *assured quality routing* (AQR). AQR, a term coined by ITSP iBasis, enables *voice over IP* (VoIP) carriers to transport voice traffic across the Internet until congestion results in unacceptable levels of quality, at which time AQR kicks in and routes calls off to the PSTN. This hybrid approach to voice is ideal for the service provider because it enables them to keep their costs down through the use of the Internet, but at the same time it enables them to guarantee quality by using the PSTN as a fallback mechanism. The customer doesn't know, nor do they need to. Once again we see the magic of the opaque cloud.

In addition to those mentioned previously, service providers must also consider the following factors when assessing the QoS level they provide:

- **Scalability**   To what degree is their technology infrastructure scalable in terms of shelf space and ports?
- **Performance**   Does the infrastructure guarantee low-packet delay and minimal jitter on an end-to-end basis?
- **Service quality**   What standards does the manufacturer claim to abide by? Are they ISO 9000 certified?
- **Reliability**   How resilient is the hardware platform? Does it guarantee *five nines* of reliability (99.999 percent uptime)?
- **Management capability**   Which *operation support system* (OSS) is used? Does it support such critical functions as billing and reconfiguration?
- **Interworking**   To what degree does the hardware/software package support the capability to effortlessly and seamlessly migrate between circuit and packet infrastructures?
- **Signaling**   Which protocols are supported? Q.931, SS7, or *multifrequency* (MF)?

SERVICE AND PROTOCOL INTEROPERABILITY

The control protocols used to maintain QoS for VoIP and other packet-based services can be divided into three primary groups.

The first of the three controls is *network-to-network interworking*. In the PSTN, this is a central issue and has historically been one of the principal considerations of telephone company design engineers. Because of the large number of service providers and manufacturers active in the industry today, it is critical that some form of network-to-network interface be defined that enables an information exchange between providers and between equipment manufactured by distinct vendors. Although this capability is well defined in the PSTN, it is not yet well defined in the IP realm. Thus, for a provider to offer true IP telephony on a universal, full-service basis, the network-to-network control protocols must be clearly defined and distributed among all players.

The second control is *device-to-device communications*, sometimes referred to as *peer-to-peer communications*. In most cases, the network assumes that in an IP environment, client devices will have some level of intelligence and the capability to establish sessions with other devices on an end-to-end basis, either directly or by first establishing a session with some sort of mediation device, such as a gateway or gatekeeper. Because these devices are intelligent, they can provide enhanced features—CLASS, multiparty calling, bandwidth management, and so on. Included here is *system-to-system messaging*, which addresses billing, directory interoperability, provisioning, and interplatform messaging. Finally, *client-to-system messaging* is important because it takes interoperability down to the customer level. Client-to-system messaging includes directory access, access to messaging applications, and client access to certain network features. A number of protocols are driving interoperability development, including the *Lightweight Directory Access Protocol* (LDAP), the *Internet Message Access*

Protocol (IMAP), the *Extensible Markup Language* (XML), the *Wireless Access Protocol* (WAP), and a number of others.

The third control is *device management*, which encompasses both the control and content functions of VoIP service. The control function hosts the intelligence required to carry out telephony functions, while the content component carries out the mandates of the control component. For the purposes of IP telephony, there must exist a mechanism to perform call control, channel specifications, and *coder/decoder* (CODEC) configurations. To a large extent, these have been addressed by protocols such as H.323, the SIP, and the MGCP.

The protocols used for all three of these functions, from the *local area network* (LAN) to the WAN and back again, are described here.

QoS MANAGEMENT IN LANs

QoS begins at the customer location and is ideally maintained through the wide area to the receiving user device. It is important, therefore, that we begin our discussion of QoS with LANs.

Because of the migration to switched architectures in the LAN environment, the *Institute of Electrical and Electronics Engineers* (IEEE) released a standard in late 1998 for the creation of *virtual LANs* (VLANs). A VLAN is nothing more than a single LAN that has been logically segmented into user groups or functional organizations. The elegance of the VLAN concept is that the members of a particular VLAN do not have to be physically collocated on the same network segment. They can be scattered all over the world as long as they are part of the same corporate network.

QoS is necessary in the LAN environment to overcome the disparity that often exists between the bandwidth offered by the LAN and the capacity of the WAN that it communicates with. LAN bandwidth has always been greater than that of the WAN, and with the arrival of Fast and Gigabit Ethernet, the discrepancy between the two is widening at a rapid rate. Furthermore, the volume of LAN traffic is growing quickly, which means that the number of LANs contending for scarce WAN bandwidth is growing. Additionally, growth in IP-based networks, especially VPNs, is forcing that demand upward.

Several IEEE standards address the concerns of LAN QoS: 802.1Q and 802.1p. Both are discussed in this section.

802.1Q The 802.1Q standard defines the interoperability requirements for vendors of LAN equipment wanting to offer VLAN capabilities. The standard was crafted to simplify the automation, configuration, and management of VLANs, regardless of the switch or end-station vendor.

Like *multiprotocol label switching* (MPLS), 802.1Q relies on the use of priority tags, which indicate service classes within the LAN. These tags form part of the frame header and use three bits to uniquely identify seven service classes. These classes, as proposed by the IEEE, are shown in Table 3-1.

802.1P Closely associated with 802.1Q is 802.1p,[3] which enables the three QoS bits in the 802.1Q VLAN header to specify QoS requirements. It is primarily used by layer 2 bridges to filter and prioritize multicast traffic. The 802.1p QoS bits can

TABLE 3-1   LAN Service Classes

PRIORITY	BINARY VALUE	TRAFFIC TYPE
7	111	Network control
6	110	Interactive voice
5	101	Interactive multimedia
4	100	Streaming multimedia
3	011	Excellent effort
2	010	Spare
1	001	Background

[3]The fact that the Q is uppercase and the p is lowercase is not an accident. IEEE 802.1p is an adjunct standard and does not stand on its own. IEEE 802.1Q, on the other hand, is an independent standard.

be set by intelligence in the client machine, as dictated by network policy established by the network management organization. In a practical application, 802.1p can be converted to DiffServ for QoS integration across the wide area. After all, 802.1p is really a QoS specification for LAN environments, most typically Ethernet. Therefore, the DiffServ byte in the IP header can be encoded at the edge of the network by the ingress router, based on information contained in the 802.1p field in the Ethernet frame header. At the egress router, the opposite occurs, guaranteeing end-to-end QoS across the wide area.

Of course, these standards only address the requirements of LANs, which will inevitably interconnect with WAN protocols, such as ATM or IP. Consequently, the *Internet Engineering Task Force* (IETF) DiffServ committee has developed standards for interoperability between 802.1Q and wide area protocols, such as IP's DiffServ, while the ATM Forum has a similar effort underway to map 802.1Q priority levels to ATM service classes.

DIFFSERV AND MPLS

The IETF has taken an active role in the development of QoS standards for IP-based transmissions. Under their auspices, two working groups have emerged with responsibilities for QoS issues. The first is DiffServ; the second is the MPLS Working Group.

When establishing connections for VoIP, it is critical to manage queues to ensure the proper treatment of packets that derive from delay-sensitive services. In order to do this, packets must be differentiable; that is, voice and video packets must be identifiable so that they can be treated properly. Routers, in turn, need to be able to respond properly to delay-sensitive traffic by implementing queue management processes. This requires the routers to establish both normal and expedited queues and to handle traffic in expedited routing queues faster than the arrival rate of the traffic. This translates into a traffic-policing requirement to ensure that the offered load remains below the bandwidth reserved at each node for high-priority data.

DIFFERENTIAL SERVICE (DIFFSERV)   DiffServ and MPLS have the same goal in mind, but approach it from different directions. DiffServ has the capability to prioritize packets through the use of bits in the IP header known as the *differential services code point* (DSCP), formerly part of the *Type of Service* (TOS) field. It relies on *per hop behaviors* (PHBs), which define the traffic characteristics that must be accommodated. The best known of these is the Expedited Forwarding PHB, designed to be used for services that require minimum delay and jitter, such as voice and video. DiffServ, then, is a technique for classifying packets according to QoS requirements. Because the classification process occurs at the edge of the network, it scales well as the network grows.

DiffServ breaks the responsibilities of traffic management into four key areas, based on the overall architecture of the network, as illustrated in Figure 3-4. At the customer's access router, traffic is managed according to flow requirements and

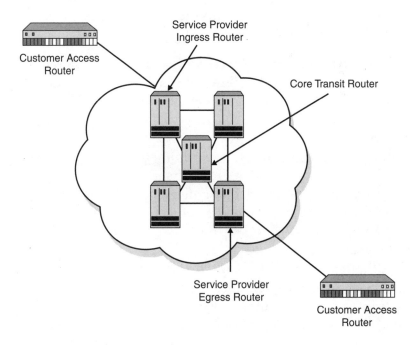

FIGURE 3-4   A DiffServ network.

is clustered for delivery to the service provider's network. The traffic is then handed to the service provider's ingress router, which sits at the edge of the network and is responsible for implementing the SLA between the customer and the service provider.

Once the edge routers have classified the traffic, the core routers can handle it according to the DSCP markers they assign to each packet. Within the network, core transit routers simply route the traffic as required. By the time the traffic arrives at the core, the edge devices have already classified it, and the core router simply handles the interior routing function. Ultimately, the traffic reaches the service provider's egress router, another edge device, which performs additional traffic-shaping functions to ensure compliance with the SLA. Thus, the core can be extremely fast because the classification process has already been done at the point of ingress.

DiffServ therefore is *not* an end-to-end protocol. Traffic, perhaps from a LAN, arrives at the edge of the network, where DiffServ's domain begins. The ingress and egress routers manage and shape traffic flows, with the freedom to use packet discards if appropriate.

MULTIPROTOCOL LABEL SWITCHING (MPLS)   MPLS is considered to be superior to DiffServ, although both techniques rely on edge routers to classify and label the packets at the point of ingress. MPLS achieves the same goal as DiffServ by establishing virtual circuits known as *label switched paths* (LSPs), which are built around specific QoS requirements. Thus, a router can establish LSPs with explicit QoS capabilities and route packets to those LSPs as required, guaranteeing the delay that a particular flow will encounter end-to-end. Some industry analysts have compared MPLS LSPs to the trunks that are established in the voice environment.

Although both MPLS and DiffServ offer reasonable levels of QoS control in IP environments, neither has the level of capability that ATM offers, which explains why ATM combined with IP is the technology of choice for many players looking to deploy mixed services over IP. Both techniques offer services to the user, but they are lacking in a number of key QoS areas, such as routing capability. Both DiffServ and MPLS describe techniques for classifying and labeling a variety of QoS levels, but neither of them speak to the requirements of establishing an end-to-end path that offers constant quality. So although both (and especially MPLS for its simplicity) are often compared to ATM, ATM is clearly a more robust and capable protocol.

Multiprotocol over ATM (MPOA) is one technique that ATM uses to manage ingress traffic and identify flows that require diverse QoS levels. MPOA establishes virtual channels for specific QoS between various devices within the network, and routes traffic as befits its unique requirements for service. If it finds traffic that requires specific handling that does not already have a virtual channel established, it creates a default channel to ensure delivery. Thus, ATM handles both packet classification *and* routing.

MPLS is similar to MPOA and uses a two-part process for routing. First, it divides the packets into various *forwarding equivalence classes* (FECs) and then maps the FECs to their next hop point. This process is performed at the point of ingress. Each FEC is given a fixed-length label that accompanies each packet from hop to hop. At each router, the FEC label is examined and used to route the packet to the next hop point where it is assigned a new label.

INTEGRATED SERVICE (INTSERV) AND THE RESOURCE RESERVATION PROTOCOL (RSVP)

Although *integrated services* (IntServ) and RSVP seem to be waning in popularity as broadly accepted QoS techniques, they are worth discussing.

IntServ is primarily seen as a useful QoS mechanism for the edge of the network and is divided into two pieces. The first piece consists of devices that have the capability to differentiate between various QoS levels so that data flows with differing requirements can be properly handled per the stipulations of the SLA. The second piece is a notification mechanism that enables an application requesting service to set up a path that will, in fact, guarantee that service on an end-to-end basis.

HOW RSVP WORKS When a host requests a specific QoS level for its data stream, RSVP transports the setup request through the network on a node-by-node basis. At each router, the RSVP setup packet attempts to reserve resources if they are available in order to maintain the QoS state until the transmission is complete. What arrives at the receiver is a rather large packet that contains path information that describes the QoS that can be expected from that path. The receiving application examines the information contained in the RSVP setup packet, requests performance parameters, and hands the information back to the RSVP, which transmits the information upstream to the originating device. The device hands the packet to the sending application, notifying it of the level of service it can expect across the requested path. The routers along the way make every effort to satisfy the request for service.

RSVP, then, is a flow-based protocol that enables an application to communicate its QoS requirements on an end-to-end basis, stipulating the bandwidth, latency, and jitter requirements that must be met for the application at hand.

A primary assumption made by RSVP is that it will largely be used for video and other high-speed multicast applications. It therefore attempts to inject a modicum of determinism in a connectionless network so that multimedia streams can be transported with acceptable levels of transmission quality.

RSVP's greatest shortcoming is its lack of scalability. It performs well when implemented in small networks, but falls short when used in high-volume, wide area situations. In order for it to work properly, RSVP requires the establishment of end-to-end paths with guaranteed quality. In a small, private network, this can be done, but when the Internet is the WAN, it is simply not possible to maintain a guaranteed end-to-end quality state today. RSVP also places a significant processing load on routers, which can cause service quality to be downgraded.

In summary, RSVP is not a routing protocol. That function is performed by other protocols such as IP, *Open Shortest Path First* (OSPF), the *Routing Information Protocol* (RIP), the

Border Gateway Protocol 4 (BGP4), and *Enhanced Interior Gateway Routing Protocol* (EIGRP). Its sole responsibility is to ensure that resources are available for a particular QoS level before enabling a session to be established that requires that level of quality.

SIGNALING IN VOIP ENVIRONMENTS

Telephony, whether circuit-switched or IP-based, requires both basic and supplementary (sometimes called enhanced) services. The basic services include call management, call authentication, caller identification, security, billing, network management, and bandwidth assignment capabilities. Supplementary services include unified messaging, integrated voice responses, 911 service, 800-number service, conference calling, online directory service, and many of the CLASSs that SS7 makes possible. In this section, we will discuss the most common signaling protocols that have been proposed for use in the IP telephony domain.

It is important to recognize that in the PSTN, SS7 enables the signaling information and the actual content to be transported across separate network infrastructures. In IP, however, the two share the same network. In the IP domain, protocols such as H.323 or SIP are responsible for signaling, including call setup and device mapping. MGCP and SS7 control the interface to gateways and call agents. The content, then, is governed by protocols such as the *Real-Time Transport Protocol* (RTP).

H.323

The protocol H.323 started as H.320 in 1996, an *International Telecommunication Union-Telecommunications Standardization Sector* (ITU-T) standard for the transmission of multimedia content over ISDN. Its original goal was to connect LAN-based multimedia systems to network-based multimedia systems. It defined a network architecture that included gatekeepers, which performed zone management and address conversion;

endpoints, which were terminals and gateway devices; and multimedia control units, which served as bridges between multimedia types.

H.323 has been rolled out in four phases. Phase one defined a three-stage call setup process: a precall step, which performed user registration, connection admission, and the exchange of status messages required for call setup; the actual call setup process, which used messages similar to ISDN's Q.931; and finally a capability exchange stage, which established a logical communications channel between the communicating devices and identified conference management details.

Phase two enabled the use of RTP over ATM, which eliminated the added redundancy of IP and also provided for privacy, authentication, and greatly demanded telephony features such as call transfer and call forwarding. RTP has an added advantage: When errors result in packet loss, RTP does not request resends of those packets, thus providing for the real-time processing of application-related content. No delays result from errors.

Phase three added the capability to transmit real-time fax after establishing a voice connection. Phase four, released in May 1999, added call connection over the *User Datagram Protocol* (UDP), which significantly reduced call setup times; inter-zone communications; call hold, park, and pickup; and call- and message-waiting features. This last phase bridged the considerable gap between IP voice and IP telephony.

One concern with H.323 is interoperability. Currently, three versions are available, and because of a lack of deployment coordination all three have a significant installed base, which naturally leads to a certain amount of fragmentation. Consequently, interoperability has become a serious concern.

One solution that has been offered is the iNOW! Profile (**www.imtc.org/act_inow.htm**). Developed initially by Lucent, VocalTec, and ITXC, iNOW! was created to define the options that manufacturers can use to ensure interoperability for multimedia voice and fax. INOW! recently merged with the *International Multimedia Teleconferencing Consortium* (IMTC). The IMTC's stated goals are to sponsor and conduct interoperability tests between suppliers of conferencing products and services, provide a forum for technical exchanges between IMTC members that will lead them to make additional submissions to the standards bodies to support the interoperability and usability of multimedia teleconferencing products and services, and finally to educate the business and consumer communities on the benefits of the underlying technologies and applications.

Under the auspices of IMTC, iNOW! has the following responsibilities:

- To help members build interoperable IP telephony products based on the current and future IMTC iNOW! interoperability profiles developed and approved by the Activity Group.
- To provide a forum for technical exchanges and to help with the resolution of technical issues that affect interoperability between iNOW!-compliant products and other products.
- To define test strategies that member companies can use to test their products for interoperability.
- To support IMTC and *European Telecommunications Standards Institute* (ETSI) TIPHON testing of cross-vendor product testing.
- To support the development of additional features and a broader scope for IMTC iNOW! profiles as the need arises.

Several Internet telephony interoperability concerns are addressed by H.323. These include gateway-to-gateway interoperability, which ensures that telephony can be accomplished between different vendors' gateways; gatekeeper-to-gatekeeper interoperability, which does the same thing for different vendors' gatekeeper devices; and finally gateway-to-gatekeeper interoperability, which completes the interoperability picture.

SESSION INITIALIZATION PROTOCOL (SIP)

Although H.323 has its share of supporters, it is slowly being edged out of the limelight by the IETF's SIP. SIP supporters claim that H.323 is far too complex and rigid to serve as a standard for basic telephony setup requirements, arguing that SIP, which is architecturally simpler and imminently extensible, is a better choice.

In reality, H.323 is an *umbrella standard* that includes (among others) H.225 for call handling, H.245 for call control, G.711 and G.721 for CODEC definitions, and T.120 for data conferencing. Originally created as a technique for transporting multimedia traffic over a LAN, gatekeeper functions have been added that enable LAN traffic and LAN capacity to be monitored so that calls are established only if adequate capacity is available on the network. Later, the Gatekeeper Routed Model was added, which enabled the gatekeeper to play an active role in the actual call setup process. This meant that H.323 had migrated from being purely a peer-to-peer protocol to having a more traditional, hierarchical design.

The greatest advantage that H.323 offers is maturity. It has been available for some time now and, although robust and full-featured, was not originally designed to serve as a peer-to-peer protocol. Its maturity therefore is not enough to carry it. It currently lacks a network-to-network interface, and it does not adequately support congestion control. This is not generally a problem for private networks, but it does become problematic for service providers who want to interconnect PSTNs and provide national service among a cluster of providers. As a result of this, many service providers have chosen to deploy SIP instead of H.323 in their national networks.

SIP is designed to establish peer-to-peer sessions between Internet routers. The protocol defines a variety of server types, including feature servers, registration servers, and redirect servers. SIP supports fully distributed services that reside in the actual user devices, and because it is based on existing IETF protocols, SIP provides a seamless integration path for voice/data integration.

Similar to H.323, SIP does not yet offer a network-to-network interface, but the IETF has created a working group designed to assemble the best features of SIP and MeGaCo (an important protocol evolving from *Media Gateway Control* that will be discussed in detail in the next section) to overcome this obstacle.

Ultimately, telecommunications, like any industry, revolves around profitability. Any protocol that enables new services to be deployed inexpensively and quickly immediately catches the eye of service providers. Like TCP/IP, SIP provides an open architecture that can be used by any vendor to develop products, thus ensuring multivendor interoperability. And because SIP has been adopted by such powerhouses as Lucent, Nortel, Cisco, Ericsson, and 3Com, and is designed for use in large carrier networks with potentially millions of ports, its success is reasonably assured.

Originally, H.323 was to be the protocol of choice to make this possible. And although H.323 is clearly a capable suite of protocols and is indeed quite good for VoIP services that derive from ISDN implementations, it is still incomplete and quite complex. As a result, it has been relegated for use as a video control protocol and for some gatekeeper-to-gatekeeper communications functions.

The intense interest in moving voice to an IP infrastructure is driven by simple and understandable factors: cost of service and enhanced flexibility. However, in keeping with the *Jurassic Park effect* (just because you *can* doesn't necessarily mean you *should*), it is critical to understand the differences that exist between simple voice and full-blown telephony with its many enhanced features. It is the feature set that gives voice its range of capabilities. A typical local switch such as Lucent Technologies' 5ESS offers more than 3,000 features, and more will certainly follow. Of course, the features and services are possible because of the protocols that have been developed to provide them across an IP infrastructure.

MEDIA GATEWAY CONTROL PROTOCOL (MGCP) AND FRIENDS

Many of the protocols that are guiding the successful development of VoIP efforts today stem from work performed early on by Level 3 and Telcordia, which together founded an organization called the International SoftSwitch Consortium. In 1998,

Level 3 brought together a collection of vendors who collaboratively developed and released the *Internet Protocol Device Control* (IPDC). At the same time, Telcordia created and released the *Simple Gateway Control Protocol* (SGCP). The two were later merged to form the MGCP, discussed in detail in RFC 2705.

MGCP enables a network device responsible for establishing calls to control the devices that actually perform IP voice streaming. It permits software call agents and media gateway controllers to control streaming media gateways at the edge of the network. These gateways can be cable modems, set-top boxes, PBXs, *Voice and Telephony over ATM* (VTOA), or VoIP. Under this design, the gateways manage the circuit-switch-to-IP voice conversion, while the agents manage signaling and call processing.

MGCP makes the assumption that call control in the network is software based, resident in external intelligent devices that perform all call control functions. It also makes the assumption that these devices will communicate with one another in a primary-secondary arrangement, under which the call agents send instructions to the gateways for execution.

Meanwhile, Lucent created a new protocol called the *Media Device Control Protocol* (MDCP), and the best features of the original three were combined to create a full-featured protocol called the MeGaCo, also defined as H.248. In March 1999, the IETF and ITU met collaboratively and created a formal technical agreement between the two organizations, which resulted in a single protocol with two names. The IETF calls it MeGaCo; the ITU calls it H.GCP.

MeGaCo/H.GCP operates under the assumption that network intelligence is housed in the central office and therefore replaces the gatekeeper concept proposed by H.323. By managing multiple gateways within a single IP-equipped central office, MeGaCo minimizes the complexity of the telephone network. In other words, a corporation might be connected to an IP-capable central office, but because of the IP-capable switches in the central office (a 7 R/E or Succession, for example) that have the capability to convert between circuit-switched and packet-switched voice, full telephony features are possible. Thus, the next-generation switch converts between circuit and packet, while MeGaCo performs the signaling necessary to establish a call across an IP WAN. It effectively bridges the gap between legacy SS7 signaling and the new requirements of IP, and supports both connection-oriented and connectionless services.

The applications that reside within networks have varying requirements with regard to bandwidth and latency. Table 3-2 summarizes them and compares the various application characteristics. Needless to say, their signaling requirements are equally varied.

IP TELEPHONY SUMMARY

Let's summarize how a telephone call is carried across an IP network. A customer begins the call using a traditional telephone. The call is carried across the PSTN to an IP telephony gateway, which is nothing more than a special purpose router designed to interface between the PSTN and a packet-based IP network. As soon as the gateway receives the call, it interrogates an associated gatekeeper device, which provides information about billing, authorization, authentication, and call routing. As soon as the gatekeeper has delivered the information to the gateway, the gateway transmits the call across the IP network to another gateway, which in turn hands the call off to the local PSTN at the receiving end, completing the call.

TABLE 3-2   Network Application Requirements

APPLICATION	REQUIRED BANDWIDTH	SENSITIVITY TO DELAY
Voice	Low	High
Video	High	Medium to High
Medical Imaging	Medium to High	Low
Web Surfing	Medium	Medium
LAN Interconnection	Low to High	Low
E-mail	Low	Low

IP telephony is primarily served by two protocols, H.323 and MGCP. H.323, from the ITU-T, is a peer-to-peer protocol under which the station initiating the call and the station terminating the call are peers with one another. The protocol often defines a gatekeeper responsible for address translation and zone management, but at the same time requires that gateway devices and terminals be responsible for call control and call-processing functions. As a call control protocol, H.323 is most capable when implemented in an environment in which the end stations are intelligent, such as PC telephony client devices and integrated PBX/gateways. Because of its complexity, however, H.323 is being pressured by the far simpler SIP protocol.

MGCP, on the other hand, is an IETF protocol born of the combination of Telcordia's SGCP and IPDC. Originally designed for use in gateway devices, it is now often found in client devices, and although designed for telephony, also serves well as a multimedia control protocol.

VOIP IMPLEMENTATION SUMMARY

There is no question that IP voice has become a reasonable technological alternative to traditional PCM voice. VoIP gateways are in various stages of development with regard to reliability, features, and manageability. Consequently, service providers interested in deploying VoIP solutions have a number of possible options available to them. One is to accept the current state of the technology and deploy it today in trial mode while waiting for enhanced versions to arrive. This would gain them an early lead, but would probably be a problem in that it would lock them into inferior technology that would ultimately set them back relative to their competitors.

A second option is a variation of the first: Implement the technology that is available today, but make no guarantees of service quality. This approach is currently being used in Western Europe and in the Southwestern United States as a way to provide low-cost long-distance service to guest worker communities. The service is actually quite good, and although it is not toll quality, it is inexpensive. And because the marketers of the service make it clear at the time of the sale that the quality will often not be equivalent to that of the traditional telephone network, there is no danger of misrepresentation. Furthermore, the companies deploying the service have no billing infrastructure to support because they rely exclusively on prepaid calling cards; thus, they can keep their costs low.

Although many believe that IP telephony will fail due to its perceived lower-service quality, others point to cellular telephony, the success of which has far exceeded all expectations in spite of the inferior quality that it delivers. Clearly, then, the market will accept lower than toll-quality voice if there is perceived value in the tradeoff.

VoIP faces a number of challenges in the marketplace today. These include the proprietary nature of gatekeeper software, the fact that many devices only offer partial compliance with H.323, a proliferation of competing standards bodies (including the IETF, the ITU, ETSI's TIPHON, and the IMTC), and the actual implementation of the technology (demonstrating VoIP's capabilities with working customers under realistic load conditions).

Another challenge is the fact that many VoIP solutions today require the user to dial long strings of digits before their call can be connected. This does not represent a simplification of telephony due to the implementation of IP.

And the bottom line? Ultimately, IP's success, whether for voice or data, hinges on one clear fact: It will only be widely accepted if it has the capability to offer customers a true VPN that delivers all the services requested over a single network interface with the expected high level of quality.

When implementing VoIP technology as a migration path, corporations should be sensitive to such obvious concerns as QoS, billing, network management, and security. But they should also question the efficacy of issues such as short- and long-term costs, scalability, standards compliance, and interoperability.

NETWORK MANAGEMENT FOR QOS

Because of the diverse audiences that require network performance information and the importance of SLAs, the data collected by network management systems must be malleable so that it can be formatted for different sets of corporate eyes. For the purposes of monitoring performance relative to SLAs, customers require information that details the performance of the network relative to the requirements of their applications. For network operations personnel, reports must be generated that detail network performance to ensure that the network is meeting the requirements of the SLAs that exist between the service provider and the customer. Finally, for the needs of sales and marketing organizations, reports must be available that enable them to properly represent the company's abilities to customers and that enable them to anticipate requirements for network augmentation and growth.

For the longest time, the *Telecommunications Management Network* (TMN) has been considered the ideal model for network management. As the network profile has changed, however, with the steady migration to IP and a renewed focus on service rather than technology, the standard TMN philosophy has begun to appear somewhat tarnished.

Originally designed by the ITU-T, TMN is built around the *Open Systems Interconnection* (OSI) model and its attendant standards, which include the *Common Management Information Protocol* (CMIP) and the *Guidelines for the Development of Managed Objects* (GDMO).

TMN employs a model, shown in Figure 3-5, comprising a *network element layer,* an *element management layer*, a *network management layer*, a *service management layer*, and a *business management layer*. Each has a specific set of responsibilities closely related to those of the layers that surround it.

The network element layer defines each manageable element in the network on a device-by-device basis. Thus, the manageable characteristics of each device in the network are defined at this functional layer.

FIGURE 3-5 The TMN layered model.

The element management layer manages the characteristics of the elements defined by the network element layer. The information found here includes activity log data for each element. This layer houses the actual element management systems responsible for the management of each device or set of devices in the network.

The network management layer has the capability to monitor the entire network based upon information provided by the element management layer.

The service management layer responds to information provided by the network management layer to deliver such service functions as accounting, provisioning, fault management, configuration, and security services.

Finally, the business management layer manages the applications and processes that provide the strategic business planning and tracking of customer interaction vehicles, such as SLAs.

OSI, while highly capable, has long been considered less efficient than IETF management standards, and in 1991, market forces began to effect a shift. That year, the Object Management Group was founded by a number of computer companies, including Sun, Hewlett-Packard, and 3Com, and together they introduced the *Common Object Request Broker Architecture* (CORBA). CORBA is designed to be vendor-independent and built around object-oriented applications. It enables disparate applications to communicate with each other, regardless of physical location or vendor. Although CORBA did not achieve immediate success, it has now been widely accepted, resulting in CORBA-based development efforts among network management system vendors. This may seem to fly in the face of the OSI-centric TMN architecture, but it really doesn't. TMN is more of a philosophical approach to network management and does not specify technological implementation requirements. Thus, a conversion from CMIP to the *Simple Network Management Protocol* (SNMP), or the implementation of CORBA, does not affect the overall goal of the TMN.

IMPLEMENTING IP SERVICES

So what does this all mean? Consider the directions that major manufacturers are taking today with regard to IP migration. Lucent Technologies and Ericsson, both critical players in the IP telephony marketplace, have targeted carriers with new gateway products that use *automatic number identification* (ANI) to select IP voice and eliminate the secondary dial tone that characterized early VoIP systems. Nortel Networks offer single-stage dialing (no secondary dial tone) and embedded CLASSs. Some vendors have offered customers the ability to select IP voice by simply dialing an access code: dial nine for a standard outside line; seven for a voice over IP line. Lucent Technologies, working with VocalTec, has developed interoperable gateway devices, fully compatible with the mandates of the iNOW! Profile. Motorola, working closely with NetSpeak, offers wireless IP voice services, and others will certainly follow.

It would appear that IP convergence will manifest itself in six significant ways: carrier-class IP voice, the growth of ITSPs, the expanded use of private corporate gateways, voice-enabled web sites, IP-enabled call centers, and ASPs.

CARRIER-CLASS IP VOICE

According to International Data Corporation, the domestic revenues for VoIP will grow at a compound annual rate of 103.4 percent between 1997 and 2002 to $24.39 billion. Internationally, growth will reach 100.9 percent in the same period, cresting at $20.49 billion in revenues.

Corporations that are considering a conversion to VoIP should take into account the following concerns when developing their migration strategies:

- Because standards compliance remains fragmented in the VoIP domain, corporations must be careful when selecting hardware and software platforms to ensure that they do not lock themselves into a proprietary solution. Similarly, implementers must ensure that vendor-to-vendor interoperability is assured because most IT organizations are loath to implement a single-vendor solution.

- The requirements placed upon voice networks and data networks are unique. As they converge to a common infrastructure, steps should be taken to ensure that the requirements of both can be satisfied by the new fabric. Related to this is the requirement for capable management solutions; to guarantee the QoS that negotiated SLAs demand, network managers must have access to a capable network management system that reflects the appropriate information.

- From a customer's perspective, the principal advantages of VoIP include consolidated voice, data, and multimedia transports; the elimination of parallel systems; the capability to exercise PSTN fallback in the event that the IP network becomes overly congested; and the reduction of long-distance voice and fax costs.

For an ISP or CLEC, the advantages are different, but no less dramatic, including the efficient use of network investments due to traffic consolidation, new revenue sources from existing clients because of the demand for service-oriented applications that benefit from being offered over an IP infrastructure, and the option of transaction-based billing. These can collectively be reduced to the general category of customer service, which service providers such as ISPs and CLECs should be focused on. The challenge they face will be to prove that the service quality they provide over their IP networks will be identical to that provided over traditional circuit-switched technology.

Major carriers are voting on IP with their own wallets, a sure sign of confidence in the technology. In June 1999, Level 3 became the first carrier to offer voice and data services across an all-IP network using Lucent's software-based SoftSwitch switching platform. Shortly thereafter, Frontier Corporation announced an agreement with Lucent Technologies worth several hundred million dollars to use the same technology in the core of their all-IP network. By the end of 2002, the company plans to shift $3 billion in service revenues to the IP backbone. Frontier Corporation also expects to see capital equipment costs go down by as much as 50 percent and equipment space and power costs reduced by as much as 75 percent. The SoftSwitch platform that both companies use also offers integrated operations support and application support for billing, AIN-like service creation, and customer support. These IP-based software switches enable call-processing and switching functions to be distributed across a series of integrated servers that enable an extremely fine control over availability and scalability.

Lucent Technologies and Nortel Networks have entered the fray with their own versions of *decomposed switch platforms*, offering their 7 R/E and Succession Network products, respectively. Both comprise a collection of high-end servers interconnected via IP and/or ATM that facilitates the evolution from today's narrowband network to a broadband platform that offers distributed access and control to deliver a wide variety of integrated services. In effect, these devices migrate central office functions from the core to the edge of the network.

This distribution process dramatically lowers the cost of entry and adds enormous flexibility to the service provider's capabilities by providing a bridge between the circuit- and packet-switched domains. Ultimately, these devices are designed to not only distribute the switching function, but also to reduce the number of network elements required to provide switched services, to ensure interoperability and service flexibility, to guarantee scalability, and to deliver carrier-class voice with five nines of reliability.

As we observed earlier, the key to IP's success in the voice-provisioning arena lies with its invisibility. If done correctly, service providers can add IP to their networks and maintain service quality while dramatically improving their overall efficiency. IP voice (not to be confused with Internet voice) will be implemented by carriers and corporations as a way to reduce costs and move to a multiservice network fabric. Lucent, Nortel, and Cisco have all added SS7 and IP voice capabilities to their router products and access devices, recognizing that their primary customers are looking for IP solutions. *Digital Switch Corporation* (DSC) has demonstrated AIN applications in an IP setting, and as a demonstration of its commitment to voice as a strategic payload, Cisco purchased Summa Four, the manufacturer of an IP voice switch capable of handling as many as 1,000 simultaneous IP voice calls. Qwest and Level 3 are fully committed to IP, and equipment manufacturers are providing the hardware necessary to meet their requirements. Thus, voice delivered across an IP infrastructure is a desirable component of the modern service provider's arsenal.

INTERNET TELEPHONY SERVICE PROVIDERS (ITSPS) AND PRIVATE CORPORATE GATEWAYS

Internet telephony offers revenue opportunities for backbone ISPs and legacy circuit-switched service providers that want to either enter the voice business or reduce their costs of providing service. Because of the inherently efficient nature of packet-based transmissions, IP telephony represents a real opportunity for service providers to dramatically increase the capacity of their voice networks.

However, IP telephony has a downside that cannot be ignored. A significant difference exists between running an ISP and running a telephone company, just as a difference exists between VoIP and IP telephony. Although Internet telephony is IP telephony, IP telephony is *not* necessarily Internet telephony. Many corporations have built internal IP networks that carry multimedia traffic, including voice. Because the networks are privately owned and operated, the IT staff responsible for service quality can regulate traffic across the backbone, employ proprietary technologies and products to ensure QoS, rely on PSTN fallback if appropriate, and manage user expectations to a large extent. Thus, IP can provide toll-quality service if deployed properly.

That model, however, is a far cry from Internet telephony. Today, the Internet can certainly deliver IP voice, but it cannot be relied upon to deliver sustainable QoS. Thus, any attempt in the public domain to replace circuit-switched voice with Internet telephony is doomed to face serious marketplace resistance because of the (correct) perception of inferior QoS. Until QoS protocols such as those discussed earlier become widespread and fully dependable, Internet telephony will be forced to coexist with circuit switching, which will remain viable for some time to come. Under certain circumstances, however, corporate IP gateways for voice are quite viable and are widely used today.

CLEARINGHOUSE SERVICES

An emerging aspect of IP telephony is the concept of *clearinghouse services*. ITSPs can form relationships with clearinghouses, which save ITSPs the effort and expense associated with forming individual relationships with the myriad other ITSPs they must interface with to be able to handoff calls in all areas where they want to provide service. Instead, the clearinghouse performs this function for the ITSP. The ITSP establishes a single interface with the clearinghouse, which then manages all call routing, authorization, billing, and network management for the ITSP. ITXC provides a clearinghouse service, as do AT&T's Global Clearinghouse and Arbinet.

Some ITSPs also employ postpaid calling cards; instead of the customer purchasing a prepaid card, the customer provides the service provider with credit card information that calls are billed to, once the customer provides the appropriate PIN. Thus, the billing is done after the call. Clearinghouse services manage this process as well.

IP-ENABLED CALL CENTERS

Ultimately, call centers are nothing more than enormous routers. They receive incoming data delivered using a variety of media (phone calls, e-mail messages, fax transmissions, or mail orders) and make decisions about handling them based on information contained in each message.

One challenge that has always faced call center management is the ability to integrate message types and route them to a particular agent based on specified decision-making criteria such as name, address, telephone number, e-mail address, ANI triggers, product purchase history, geographic location, or language preference. This has resulted in the development of a technical philosophy called *unified messaging*. With unified messaging, all incoming messages for a particular agent, regardless of the media over which they are delivered, are housed centrally and clustered under a single account identifier. When the agent logs into the network, their PC lists all the messages that have been received for them, giving them the ability to much more effectively manage the information contained in those messages.

Today, unified messaging systems also support the road warriors. A traveling employee can dial into a message gateway and download all messages—voice, fax, and e-mail—from that one central location, thus dramatically simplifying the process of staying connected while away from the office.

Call centers are undergoing a tremendous change as the IP juggernaut hits them. The first of these is a redefinition of the market they serve. For the longest time, call centers have primarily served large corporations because they are expensive to deploy. With the arrival of IP, however, the cost is dropping, and some industry authorities believe that by the end of 2001 the cost of a position in an IP call center will be approximately half that of a traditional call center.

The second major change is that call centers are becoming the focal point for corporations with effective and successful e-commerce applications. Small and medium-size businesses are enjoying an expanded market presence thanks to e-commerce. The Internet and the disintermediation that it brings about are helping them lower their operational costs.

One customer complaint that is often voiced about the Web, particularly with regard to e-commerce, has to do with its lack of interactivity. When customers use the Web to make purchases, they are often faced with the dilemma of needing more information than the company's web site gives them, which often requires that they disconnect from their ISP and call the company's 800 number—a step that effectively obviates the need for ordering online. In response to this, some e-commerce providers have implemented the capability to invoke an Internet telephony session with an agent while product surfing. Although the capability is not yet as good as it soon will be, the fact that major corporations are adding voice capabilities to their web sites and that customers are using the service proves that the demand for integrated voice and data must be satisfied.

INTEGRATING THE PBX

There is an enormous installed base of legacy PBX equipment, and vendors did not enter the IP game enthusiastically. Early entrants arrived with enhancements to existing equipment that were proprietary and expensive, and did very little to raise customer awareness or engender trust in the vendor's migration strategy. Over time, however, PBX manufacturers began to embrace the concept of convergence as their customers' demands for IP-based systems grew, and soon products began to appear. Most have heard the message delivered by the customer: Preserve the embedded base to the degree possible as a way of saving the existing investment, create a migration strategy that is seamless and as transparent as possible, and preserve the features and functions that customers are already familiar with to minimize churn, the tendency of customers to constantly change service providers.

Major vendors like Lucent Technologies and Nortel Networks have responded with products that do exactly that. Lucent's DEFINITY IP Solutions package is an upgrade to their DEFINITY Enterprise Communications Server PBX, which incorporates IP telephony capability. It enables voice calls and faxes to be carried over LANs, WANs, the Internet, and corporate intranets. The DEFINITY IP Solutions package functions as both a gateway and a gatekeeper, providing circuit-to-packet conversion, security, and access to a wide variety of applications; these include enhanced call features such as multiple line appearances, hunt groups, multiparty conferencing, call forwarding, holds, call transfers, and speed dial. It also provides access to voice mail, CTI applications, wireless interfaces, and call center features. An additional product, the DEFINITY IP SoftPhone, enables a PC to serve as a full-feature business telephone.

Nortel's Meridian Integrated IP Telephony Gateway is a card-based adjunct that is installed in the Meridian 1 Intelligent Peripheral Equipment shelf. It interconnects multiple systems across a private network, creating a VPN-like environment that can carry compressed voice and fax messages as IP packets across an IP network. The product is H.323-compliant and enables corporations to take advantage of Meridian's well-known feature set, including least-cost routing,

QoS discrimination, systems management, and discrete billing. It also supports the capability to fall back to the PSTN if QoS decays due to congestion or packet loss.

Like Lucent's DEFINITY, the Meridian system supports the capability to carry IP telephony down to the desktop level. The Meridian IP Telecommuter software enhancement enables a PC to serve as a full-feature telephone set.

BILLING AS A CRITICAL SERVICE

One area that is often overlooked when companies attempt to improve the quality of the services they provide to their customers is billing. Although it is not typically viewed as a strategic competitive advantage, studies have shown that customers view it as one of the top considerations when assessing a service provider.

Billing offers the potential to strengthen customer relationships, improve long-term business health, cement customer loyalty, and generally make businesses more competitive. However, for billing to achieve its maximum benefit and strategic value, it must be fully integrated with a company's other operations' support systems, including network and service provisioning systems, installation support, repair, network management, and sales and marketing. If done properly, the billing system becomes an integral component of a service suite that enables the service provider to quickly and efficiently introduce new and improved services in logical bundles. The billing system also enables the service provider to improve business indicators such as service timeliness, billing accuracy, and cost; offer custom service programs to individual customers based on individual service profiles; and transparently migrate from legacy service platforms to the so-called next-generation network.

In order for billing to work successfully as a strategic service, service providers must build a business plan and a migration strategy that takes into account various factors. These include integration with existing operations' support systems, business process interaction, the role of IT personnel and processes, and post-implementation testing to ensure compliance with strategic goals stipulated at the beginning of the project.

SUMMARY

IP is here to stay and is profoundly changing the nature of telecommunications at its most fundamental level. Applications for IP range from carrier-class voice that is indistinguishable from that provided by traditional circuit-switching platforms to best-effort services that carry no service quality guarantees. The interesting thing is that all the various capability levels made possible by the incursion of IP have an application in modern telecommunications and are being implemented at a rapid-fire pace.

There is still a tremendous amount of hype associated with IP services as they edge their way into the protected fiefdoms of legacy technologies. Implementers and customers alike must be wary of *brochureware* solutions and the downside of the Jurassic Park effect, which warns that just because you *can* implement an IP telephony solution doesn't necessarily mean you *should*. Hearkening back once again to the old telephone company adage, "If it ain't broke, don't fix it," buyers and implementers alike must be cautious as they plan their IP migration strategies.

IP technology offers tremendous opportunities to offer consolidated services, to make networks and the companies that operate them more efficient, to save cost, and to pass those savings on to the customer. Ultimately, however, IP's promise lies in its ubiquity and its capability to tie services together and make them part of a unified delivery strategy. The name of the game is service, and IP provides the bridge that enables service providers to jump from a technology-centric focus to a renewed focus on the services that customers care about.

REDEFINING THE SERVICE PROVIDER

BEGINNINGS

When we started this technological journey in the first part of the book, we discussed 16 characteristics of the evolving network and how they are affecting technology deployment and development, company positioning, and service provisioning. Let's go back and visit that list again in light of the relative roles of technology, companies, and services that we have discussed throughout the book.

Customers no longer want to buy technology. This is an absolute truth that cannot be ignored by service providers if they want to continue to play an active role in their own industry. Customers no longer want to buy technology because to a very large extent, they no longer care about it. Their concern is with the uses for the technology and the manner in which they will convert the technology delivered by service providers into a competitive advantage in their own markets. This represents an enormous opportunity for service providers who have the wherewithal to probe their customers' marketplaces and offer solutions based on communications technology that will help those customers position themselves competitively among their peers. To say that customers do not want to buy technology is perhaps a bit overstated; they clearly have a need for the technology behind telecommunications and therefore want to buy it. That is not, however, where their focus lies. Technology is a means to an end in the eyes of the customer, not an end unto itself. Purveyors of technology who fail to understand that will become second rate players and will be sidelines by those who *do* understand the difference.

Bandwidth has become a commodity. Strategic planners often talk about the *Strengths, Weaknesses, Opportunities, Threats* (SWOT) model—when analyzing themselves in the context of the markets they serve (see Figure 4-1). They typically begin by examining their own strengths, a natural place to start, and usually come up with a substantial list of capabilities that make the company strong. Feeling good about themselves, they often stop the analysis at this point or only give cursory coverage to the other three areas. This is a bad practice because the exercise is valuable and should be done in its entirety for maximum benefit. Failure to analyze functional weaknesses, marketplace opportunities, or competitive threats can lead to exposure.

When service providers analyze their strengths, they typically come up with the following list:

- History and company legacy
- Reputation for good service
- Strong customer base
- In-place, capable network backbone

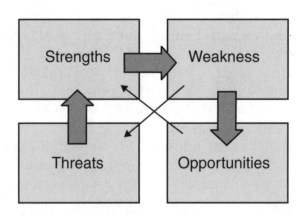

FIGURE 4-1   The SWOT model.

- People
- Size

Upon careful analysis, the list they assemble for weaknesses often includes

- History and company legacy
- Less than optimal service quality and time to market for new products and services
- Declining marketshare
- Regulatory hobbles
- Too large and therefore slow-moving

Note the various similarities with the first list. Some characteristics that were seen as strengths in the legacy world are now being viewed as liabilities. This obviously points to an area that deserves attention from strategic decision-makers.

Opportunities commonly listed by service providers include

- The ability to sell to the competition
- The opportunity to manage the competitors' networks for them as an outsource company
- The ability to offer world-class provisioning services
- Opportunities to take advantage of new technologies in the network, such as *Internet Protocol* (IP), wireless, and *Digital Subscriber Line* (DSL), to provide enhanced services
- Becoming an ISP and taking advantage of the in-place network and its surrounding capabilities

Finally, threats commonly voiced include

- Loss of marketshare to faster moving, more aggressive competitors
- Increasingly (in some eyes) punitive regulatory environment

- Growing competition
- Reduction of product profit margins

One of the greatest threats facing the incumbent service providers is the commoditization of bandwidth as it becomes available in volume and at near-zero prices. All of the concerns expressed in each of the four areas shown previously point to the need to move from being a commodity provider to being a true service provider in a competitive marketplace.

The SWOT model can be a valuable tool for analyzing market position and for determining strategic direction as companies move from undifferentiated commodity to specialized service. Strengths are comparatively easy to assess and once understood help to understand a company's weaknesses. Weaknesses are usually functions a company either does not perform at all or performs at a level that is perceived to be inferior to those the competition offers. Weaknesses, in turn, help to identify opportunities to either add new services or improve existing ones. Finally, threats must be managed and dealt with. They typically show themselves as competitors stealing market-share or as internal weaknesses that give the competition some other form of advantage.

There are two other relationships in the model that must be mentioned. Weaknesses typically lead to opportunities, but they also help to illustrate the presence of threats. Competitors, like animals, tend to strike at their enemy's weakest point, and those are the areas that should be identified and addressed early on.

The other relationship observes that those opportunities, if identified and acted upon, help a company shore up and add to strengths. The fact that bandwidth, formerly the greatest competitive strength of the legacy service providers, is becoming a widely traded commodity says that those providers must change the way they do business or face the classic business death spiral.

Customers are demanding more bandwidth. As bandwidth becomes cheaper and more widely available, applications are being developed that require it in larger and larger quantities.

Bandwidth must therefore become one component of a full-service provider's array of services. The bandwidth must be universally accessible, cost effective, scalable, and must provide access to a wide array of related services. Today, service providers are deploying technologies that will allow them to deliver that bandwidth such as cable modems, DSL, and high-speed wireless technologies like *local multipoint distribution service* (LMDS).

The role of the service provider has changed dramatically. There was a time when *service provider* referred to a company that sold access and transport. Today, service providers sell clouds. They have come to realize that success comes from understanding what the customer's business requirements are and offering solutions to the problems they face at competitive prices, without burdening the customer with the details of the underlying technology.

Managerial structures are changing. In lockstep with the flattening of computer networks, managerial structures are flattening as well. Corporations are shifting much of the operational and tactical decision-making responsibility to lower managerial levels, thus accelerating the pace at which decisions can be made and putting responsibility for those decisions in the hands of managers best equipped to make them. At the same time, these companies maintain a philosophical hierarchy by creating a corporate vision that the peer organizations look up to for guidance. Thus, they achieve their corporate goals in ways best suited to each sub organization but are all attuned to the same vision about where the company is going.

The networks that support the evolving corporations are also evolving. Again, as corporations move from being monolithic, vertically integrated organizations to virtual cluster companies, the networks they deploy internally must evolve with them. The need for flexibility, speed, and scalability has brought about the development of the *virtual private network* (VPN), a unique combination of public network infrastructure and secure transport protocols that delivers the best of the public and private network worlds.

The local loop is evolving. In keeping with customer demand for bandwidth and the service providers' need to satisfy that demand, service providers are deploying improved local loop technologies that deliver higher bandwidth without requiring massive physical upgrades to the outside plant. These include DSL, 56K modems, wireless local loop technologies, and cable modems.

Wireless access technologies are becoming centrally important. While bandwidth is in great demand, mobility and flexibility are equally sought after. Not only is wireless infrastructure less costly to deploy, it can also be installed much faster than its wire-based relative, providing a time-to-market advantage for companies capable of deploying services over it. For this reason alone, wireline companies are buying their way into the wireless domain as quickly as they can.

The fabric of the transport network is changing. As one executive recently observed, "This is not your mother's favorite circuit switch anymore." If service providers are to become all things to all applications, they must deploy a technology-rich cloud that is fully capable of delivering a wide variety of *quality of service* (QoS) levels, universally, at prices that customers will pay. Circuit switching will be in place for some time to come and will continue to be a central component of the world's *wide area network* (WAN) fabric. Over time, however, it will be replaced with a packet-based infrastructure and a suite of control protocols that will enable it to deliver the same QoS that circuit switching has been known for nearly 100 years.

IP is ascending to the throne of the protocol monarchy. The ascendancy of IP is inevitable. It is a global protocol, found in all network operating systems, offers a universally accepted addressing scheme, and interfaces with all network architectures. The beauty of IP is that it does not replace existing protocols but rather provides the interstitial fabric that ties them all together, enabling service providers to unify their product and service offerings under a common, full-service cloud.

The Internet will continue to be a focal point for the evolving network-centric world. Although it may not replace other networks, it will certainly compete with them for customers as QoS protocols and improved infrastructure make it capable of offering the same services delivered today by legacy networks.

Management roles are evolving, sometimes in unusual ways. This is not trendy; it is necessary. The staid roles of traditional management hierarchies no longer work in the accelerated world of IP networking and virtual corporations. Knowledge managers, once the stuff of Ray Bradbury and Arthur C. Clarke novels, are becoming key personnel in the modern, knowledge-driven corporation.

The knowledge-based corporation is a reality, and the knowledge worker is on the rise. This is a time in the telecommunications industry for finesse, not brute force. The force of raw technology is giving way to the finesse of knowledge-managed technology, as service providers learn how to take advantage of data mining and analysis techniques to help them meet customer requirements. Leadership no longer goes to the company with the best technology but rather to the company that has the *right* technology and the customer knowledge to properly apply it.

Intelligence is migrating from the center of the network out toward the user's access device. David Isenberg's prophetic paper spoke about the movement of network intelligence from the center of the service provider network to the edge of the network and beyond. Today, with the rise of IP telephony, service functions formerly performed by core network elements, such as *Signaling System Seven* (SS7) *service control points* (SCPs), *signal transport points* (STPs), and central office switch routing functions, are now being managed by intelligent client devices, in many cases owned and operated by the customer. Improvements in the speed and availability of specialized *digital signal processors* (DSPs) and application-specific chipsets have now made it possible to distribute the network intelligence, resulting in a far faster and more efficient service creation environment.

The players in the game are changing. Here is how one *incumbent local exchange carriers* (ILEC) executive characterized the industry today: "Imagine a playing field. On the playing field are three rugby players, a swimmer, a couple of lacrosse people, two basketball forwards, a javelin thrower, seven chess grand masters, a dart champion, and a team of skydivers. They are told to shake hands and go play. Today, that's what this industry feels like a lot of the time." The characterization is a good one. This industry no longer comprises a small collection of companies with well-defined responsibilities, territories, and rules that govern them. It has become something of a free-for-all, and although that is not necessarily bad, it does inject a degree of chaos that can make life difficult for those inside the industry and confusing for those who want to interact with it.

The traditional, local access and transport area (LATA)-bound market is a thing of the past. An old Asian curse begs, "May you live in interesting times." Although service providers often groused about the LATA restrictions that limited their abilities to enter new markets, they clearly understood them and knew well the rules of behavior. Today, all bets are off. As local companies are allowed into long distance, as long distance companies plunge into the local market, and as both buy cable properties and deploy wireless services, the rules that govern behavior in the marketplace are anyone's guess. "Be careful what you wish for—you might get it" could well be the defining mantra for today's telecommunications industry.

Those facts are as true in this second edition of the book as they were in the first. Gary Martin, the President of TM2, Inc., a telecommunications and management services consultancy in Clinton, New Jersey, sums these facts up nicely in a graph, shown in Figure 4-2, that plots the degree to which a company is efficient (X-axis) against the degree to which they are effective (Y-axis). Efficiency, says Martin, is a measure of doing things right. Effectiveness, on the other hand, is a measure of doing the right things. Both, he observes, are critical for success, but there are some interesting observations that can be made. Martin often counsels corporations about the degree to which they exemplify each measure and helps them move in the appropriate direction to ensure that they demonstrate high levels of both.

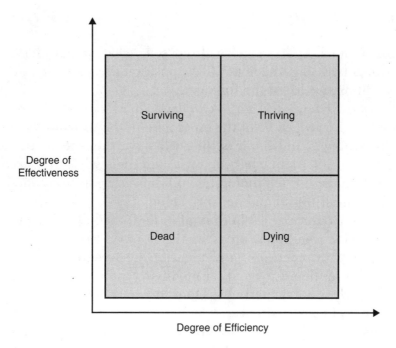

FIGURE 4-2   Efficiency versus effectiveness.

Consider each of the four quadrants of the graph. A company that is both effective and efficient, shown in the upper right quadrant, is thriving. A company that is effective but not particularly efficient (upper left quadrant) is surviving. Companies that are efficient but not particularly effective (lower right quadrant) are most likely dying—although as Martin observes, may not realize it—while companies that do neither are most likely dead or very close to it. The exercises that Martin conducts with corporate clients are often uncomfortable because they tend to bare the corporate soul and shed light upon the shortcomings of the company. The results, however, are powerful: corporate executives are given a new way of thinking about the way their companies operate as well as effective measures to chart the course through the perils of an ever-more-competitive marketplace.

MOVING FORWARD

In light of all that we have discussed, what should service providers be doing to competitively position themselves? In the first edition, we listed the following:

Manage knowledge, not technology. Think like a customer. Become knowledgeable of the environment they operate in and create products and services that will serve their needs there. Don't try to sell them what you have; sell them what they *need*. Sell solutions, not technologies, and deliver them from an all-purpose, multifunctional network cloud.

Find a parade and get in front of it. Perform SWOT analyses of all major functional areas in the company that could be attacked by competitors, and identify the weaknesses. Once found, shore them up by either modifying internal practices or allying with strategic partners. Do it for your own company; do it for that of your customers.

Build a technology-based infrastructure that is capable, robust, and future-proof. Determine the direction that your company will take and build an infrastructure that will take you there with your customers. If analyses show that IP is the right technology, then by all means make it part of the strategic technology vision of the company. Do not, however, succumb to hype and brochureware, and heed the teachings of the Jurassic Park Effect.

Think of your company as if it were two companies: a backbone provider and a professional services provider. You may not be able to physically restructure the corporation, but you can view it as if it were restructured. The backbone company provides the necessary infrastructure, while the professional services company creates the services that will be delivered over the backbone. This imaginary structure forces a gap between the technology and the services, which is crucial: They are *not* the same. One is a vehicle; the other is the service that rides over the vehicle. Both are necessary, but independent of each other.

Never forget what got your company where it is in the first place. There was a time when data was considered to be the killer application for voice networks because it could ride free atop the highly lucrative voice services. Today, the opposite is becoming true. Voice is becoming the killer application for data networks because once packetized, it can ride for free atop the data stream.

Build an independent knowledge management organization. Make them responsible for data mining, customer relationship management, knowledge management, and by extension, enterprise resource planning. Support the need for knowledge sharing at the highest management levels in the company, and make it clear that their efforts are central to the success of the corporation.

Now, in this second edition, I add the following thoughts. There are really two key actions that must be taken to preserve the long-term health of the technology and communications sectors. The first is a general overhaul of the regulatory structure; the other is to refocus the various segments of the industry.

The following steps should be taken. First, recognize that the local loop is for all intents and purposes a natural monopoly and that this fact will not change without a significant regulatory shift. This is not a U.S. phenomenon: Canada has undergone the same wrenching evolution with minimal positive effects as has the U.K. *British Telecom* (BT) complied with the Oftel mandate to unbundle the local loop as the first step toward creating a competitive local telecomm market; months later, they are still asking where the competitors are that wanted element unbundling in the first place.

One way to approach the conundrum is to remove the subsidies on local service to begin the complex process of creating a competitive market. Next, allow the ILECs to raise their rates such that they reflect the actual cost of delivering service. Subsidies can be preserved in the 18 percent of the local loops that are high-cost or serve low-income families. Subsidy dollars can then be reallocated to pay for broadband access. The model works: In Massachusetts, Verizon has slowly raised their rates from $8 per access line to an average of $21—$2 more than the actual cost of deployment; 161 viable competitors now exist in the local market, and they hold 20 percent of the market—more than five times the national average. Furthermore, studies indicate that a $2 surcharge on qualifying local loops in the United States would generate an additional $3 billion to further the broadband effort.

The second area that requires attention is a renewed focus on market dynamics by each of the industry groups, including both manufacturers and service providers. The ILECs and *post, telephone, and telegraphs* (PTTs), for example, should focus on cost reduction efforts that can be passed on to customers; push hard for regulatory relief to enter the long distance business for Internet and high-speed data; offer premium pricing terms that mirror the plans offered by interexchange carriers (multiyear contracts with annual renegotiation terms and discounts for volume purchases); offer outsourcing services, and work hard to increase their service areas.

The *competitive local exchange carriers* (CLECs) or second carriers, on the other hand, must focus on becoming solution providers rather than purely alternative access carriers. They must partner with content companies, *application service providers* (ASPs), Web hosting services, e-mail providers, security companies, and online storage companies. They must build account teams that know the market and the companies that make up the market, and should do everything in their power to offer the best possible customer service *as measured by the customer*.

Long distance companies must also create account teams that know the market well and offer full-service solutions with seamless connectivity. Customer service must be made the number one priority, and value, not price, should be the focal point. Consider this: In 1999, an OC-3/STM-1 between the U.K. and North America cost $12 million. Today, because of optical technology and the proliferation of core bandwidth, the same circuit costs $1.8 million.

Hardware manufacturers must operate under the knowledge that technology is not their ultimate product. There is no question that technology plays a fundamental and critical role in the marketplace, but its most critical role is as a facilitator of service, not a service in its own right. Their focus should be on the direction that their customers are taking with regard to their own customers' demands, delivering products that facilitate their ability to satisfy them.

Both service providers and manufacturers must realize that although there is a glut of bandwidth in the network core, there is precious little growth at the edge, and that's where the customers dwell. The customers have the money, and they are fully prepared to throw it at whatever company is willing to give them what they want. What they want is broadband access. They don't care if it comes via DSL, wireless, or a cable modem; all they really care about is that (1) it's available, and (2) it is accompanied by a suite of applications that provide tangible value in the form of differentiation or competitive advantage.

CONCLUSION

So, what is to be learned from this? First of all, the perceived decline and fall of the great telecommunications empires is not the beginning of a decent into chaos and technological barbarism, nor does it herald the beginning of a Dark Age in the marketplace. The evidence of this lies in the fact that the market remained strong (albeit with a few speed bumps) following the horrific acts of terrorism in September. The telecomm players in all their many forms are coming back, and although their return is a slow one, it is inevitable. Now, then, is the time for them to reconsider their position and the way they will play that position when the market returns to its normal state of robustness in the near future. Here are some suggestions.

MANUFACTURERS

Focus on those products that lie closest to the company's core capability set and consider jettisoning the rest. Focus on services, solutions, and tools for the creation of competitive advantage among clients; position products as the facilitators of those tools. Place the ability to manage at the network level high on the list of offered capabilities. Also, recognize that certain key technologies—optical transport, metro geography, and wireless access—will hold sway in the catbird seat for some time to come. Pay attention to them. Finally, recognize that technology without a problem to solve is cute, but largely without utility— or marketshare—today. Consider the lessons learned from the 3G debacle that continues to rage.

SERVICE PROVIDERS

Remember that *service* means more than access and transport. Customers today are looking for a business partner that can help them better position themselves among the steadily growing field of competitors. They should focus, therefore, on expansion of their core capability set by adding functionality in the form of alliances or mergers with ASPs and content providers so that they become more than just a deliverer of dial tone. Furthermore, they should learn to chant the mantra of QoS and accept the fact that it must be defined by the customer, not by the service provider.

REGULATORS

The Dark Ages are over—it's time for the Age of Enlightenment to begin. This is a very different telecommunications industry than it was at the dawn of its creation, so the rules that governed it then, and that to a large extent still govern it today, should be carefully examined for appropriateness and discarded without emotion when they are found to *not* be appropriate. It is time to recognize that from an economic point-of-view, the local loop is a natural monopoly and should be treated as such. If, however, the access marketplace is to be competitive, then drastic measures, such as structural separation, should be undertaken to make it happen. Broadband is a requirement for economic growth under the general mandate of national security, prosperity, and longevity; steps should be taken to move its widespread deployment to a position of national attention and importance.

If these steps are undertaken, telecommunications will be able to accomplish the mandate of the Communications Act of 1934 of universal, affordable service. The mandate, however, will extend beyond telephony, beyond affordability, and beyond the borders of the United States. Universal telecommunications capability is a critical component of the worldwide effort toward a globalized economy and should therefore be treated to the attention it so rightly deserves.

Let's face it: It's *still* a whole new ballgame out there, and the customers are making the rules. The service providers, however, own the stadium, and the game is theirs to lose.

COMMON INDUSTRY ACRONYMS

AAL	ATM Adaptation Layer
AARP	AppleTalk Address Resolution Protocol
ABM	Asynchronous Balanced Mode
ABR	Available Bit Rate
AC	Alternating Current
ACD	Automatic Call Distribution
ACELP	Algebraic Code-Excited Linear Prediction
ACF	Advanced Communication Function
ACK	Acknowledgment
ACM	Address Complete Message
ACSE	Association Control Service Element
ACTLU	Activate Logical Unit
ACTPU	Activate Physical Unit
ADCCP	Advanced Data Communications Control Procedures
ADM	Add/Drop Multiplexer
ADPCM	Adaptive Differential Pulse Code Modulation
ADSL	Asymmetric Digital Subscriber Line
AFI	Authority and Format Identifier
AIN	Advanced Intelligent Network
AIS	Alarm Indication Signal
ALU	Arithmetic Logic Unit
AM	Administrative Module (Lucent 5ESS)

AM	Amplitude Modulation
AMI	Alternate Mark Inversion
AMP	Administrative Module Processor
AMPS	Advanced Mobile Phone System
ANI	Automatic Number Identification (SS7)
ANSI	American National Standards Institute
APD	Avalanche Photodiode
API	Application Programming Interface
APPC	Advanced Program-to-Program Communication
APPN	Advanced Peer-to-Peer Networking
APS	Automatic Protection Switching
ARE	All Routes Explorer (Source Route Bridging)
ARM	Asynchronous Response Mode
ARP	Address Resolution Protocol (IETF)
ARPA	Advanced Research Projects Agency
ARPANET	Advanced Research Projects Agency Network
ARQ	Automatic Repeat Request
ASCII	American Standard Code for Information Interchange
ASI	Alternate Space Inversion
ASIC	Application Specific Integrated Circuit
ASK	Amplitude Shift Keying
ASN	Abstract Syntax Notation
ASP	Application Service Provider
AT&T	American Telephone and Telegraph
ATDM	Asynchronous Time Division Multiplexing
ATM	Asynchronous Transfer Mode
ATM	Automatic Teller Machine
ATMF	ATM Forum
AU	Administrative Unit (SDH)

AUG	Administrative Unit Group (SDH)
AWG	American Wire Gauge
B8ZS	Binary 8 Zero Substitution
BANCS	Bell Administrative Network Communications System
BBN	Bolt, Beranak, and Newman
BBS	Bulletin Board Service
Bc	Committed Burst Size
BCC	Blocked Calls Cleared
BCC	Block Check Character
BCD	Blocked Calls Delayed
BCDIC	Binary Coded Decimal Interchange Code
Be	Excess Burst Size
BECN	Backward Explicit Congestion Notification
BER	Bit Error Rate
BERT	Bit Error Rate Test
BGP	Border Gateway Protocol (IETF)
BIB	Backward Indicator Bit (SS7)
B-ICI	Broadband Intercarrier Interface
BIOS	Basic Input/Output System
BIP	Bit Interleaved Parity
B-ISDN	Broadband Integrated Services Digital Network
BISYNC	Binary Synchronous Communications Protocol
BITNET	Because It's Time Network
BITS	Building Integrated Timing Supply
BLSR	Bidirectional Line Switched Ring
BOC	Bell Operating Company
BPRZ	Bipolar Return to Zero
Bps	Bits per second
BRI	Basic Rate Interface

BRITE	Basic Rate Interface Transmission Equipment
BSC	Binary Synchronous Communications
BSN	Backward Sequence Number (SS7)
BSRF	Bell System Reference Frequency
BTAM	Basic Telecommunications Access Method
BUS	Broadcast Unknown Server
C/R	Command/Response
CAD	Computer-Aided Design
CAE	Computer-Aided Engineering
CAM	Computer-Aided Manufacturing
CAP	Carrierless Amplitude/Phase modulation
CAP	Competitive Access Provider
CARICOM	Caribbean Community and Common Market
CASE	Common Application Service Element
CASE	Computer-Aided Software Engineering
CAT	Computer-Aided Tomography
CATIA	Computer-Assisted Three-dimensional Interactive Application
CATV	Community Antenna Television
CBEMA	Computer and Business Equipment Manufacturers Association
CBR	Constant Bit Rate
CBT	Computer-Based Training
CC	Cluster Controller
CCIR	International Radio Consultative Committee
CCIS	Common Channel Interoffice Signaling
CCITT	International Telegraph and Telephone Consultative Committee
CCS	Common Channel Signaling
CCS	Hundred Call Seconds per Hour
CD	Collision Detection
CD	Compact Disc

CDC	Control Data Corporation
CDMA	Code Division Multiple Access
CDPD	Cellular Digital Packet Data
CD-ROM	Compact Disc-Read Only Memory
CDVT	Cell Delay Variation Tolerance
CEI	Comparably Efficient Interconnection
CEPT	Conference of European Postal and Telecommunications Administrations
CERN	European Council for Nuclear Research
CERT	Computer Emergency Response Team
CES	Circuit Emulation Service
CEV	Controlled Environmental Vault
CGI	Common Gateway Interface (Internet)
CHAP	Challenge Handshake Authentication Protocol
CICS	Customer Information Control System
CICS/VS	Customer Information Control System/Virtual Storage
CIDR	Classless Interdomain Routing (IETF)
CIF	Cells In Frames
CIR	Committed Information Rate
CISC	Complex Instruction Set Computer
CIX	Commercial Internet Exchange
CLASS	Custom Local Area Signaling Services (Bellcore)
CLEC	Competitive Local Exchange Carrier
CLLM	Consolidated Link Layer Management
CLNP	Connectionless Network Protocol
CLNS	Connectionless Network Service
CLP	Cell Loss Priority
CM	Communications Module (Lucent 5ESS)
CMIP	Common Management Information Protocol
CMISE	Common Management Information Service Element

CMOL	CMIP Over LLC
CMOS	Complementary Metal Oxide Semiconductor
CMOT	CMIP Over TCP/IP
CMP	Communications Module Processor
CNE	Certified NetWare Engineer
CNM	Customer Network Management
CNR	Carrier-to-Noise Ratio
CO	Central Office
CoCOM	Coordinating Committee on Export Controls
CODEC	Coder-Decoder
COMC	Communications Controller
CONS	Connection-Oriented Network Service
CORBA	Common Object Request Brokered Architecture
COS	Class of Service (APPN)
COS	Corporation for Open Systems
CPE	Customer Premises Equipment
CPU	Central Processing Unit
CRC	Cyclic Redundancy Check
CRT	Cathode Ray Tube
CRV	Call Reference Value
CS	Convergence Sublayer
CSA	Carrier Serving Area
CSMA	Carrier Sense Multiple Access
CSMA/CA	Carrier Sense Multiple Access with Collision Avoidance
CSMA/CD	Carrier Sense Multiple Access with Collision Detection
CSU	Channel Service Unit
CTI	Computer Telephony Integration
CTIA	Cellular Telecommunications Industry Association
CTS	Clear To Send

CU	Control Unit
CVSD	Continuously Variable Slope Delta modulation
CWDM	Coarse Wavelength Division Multiplexing
D/A	Digital-to-Analog
DA	Destination Address
DAC	Dual Attachment Concentrator (FDDI)
DACS	Digital Access and Cross-connect System
DARPA	Defense Advanced Research Projects Agency
DAS	Dual Attachment Station (FDDI)
DASD	Direct Access Storage Device
DB	Decibel
DBS	Direct Broadcast Satellite
DC	Direct Current
DCC	Data Communications Channel (SONET)
DCE	Data Circuit-terminating Equipment
DCN	Data Communications Network
DCS	Digital Cross-connect System
DCT	Discrete Cosine Transform
DDCMP	Digital Data Communications Management Protocol (DNA)
DDD	Direct Distance Dialing
DDP	Datagram Delivery Protocol
DDS	DATAPHONE Digital Service (Sometimes Digital Data Service)
DDS	Digital Data Service
DE	Discard Eligibility (LAPF)
DECT	Digital European Cordless Telephone
DES	Data Encryption Standard (NIST)
DID	Direct Inward Dialing
DIP	Dual Inline Package
DLC	Digital Loop Carrier

DLCI	Data Link Connection Identifier
DLE	Data Link Escape
DLSw	Data Link Switching
DM	Delta Modulation
DM	Disconnected Mode
DMA	Direct Memory Access (computers)
DMAC	Direct Memory Access Control
DME	Distributed Management Environment
DMS	Digital Multiplex Switch
DNA	Digital Network Architecture
DNIC	Data Network Identification Code (X.121)
DNIS	Dialed Number Identification Service
DNS	Domain Name System (IETF)
DOD	Direct Outward Dialing
DoD	Department of Defense
DoJ	Department of Justice
DOV	Data Over Voice
DPSK	Differential Phase Shift Keying
DQDB	Distributed Queue Dual Bus
DRAM	Dynamic Random Access Memory
DSAP	Destination Service Access Point
DSF	Dispersion-Shifted Fiber
DSI	Digital Speech Interpolation
DSL	Digital Subscriber Line
DSLAM	Digital Subscriber Line Access Multiplexer
DSP	Digital Signal Processing
DSR	Data Set Ready
DSS	Digital Satellite System
DSS	Digital Subscriber Signaling System

DSU	Data Service Unit
DTE	Data Terminal Equipment
DTMF	Dual Tone Multifrequency
DTR	Data Terminal Ready
DVRN	Dense Virtual Routed Networking (Crescent)
DWDM	Dense Wavelength Division Multiplexing
DXI	Data Exchange Interface
E/O	Electrical-to-Optical
EBCDIC	Extended Binary Coded Decimal Interchange Code
ECMA	European Computer Manufacturer Association
ECN	Explicit Congestion Notification
ECSA	Exchange Carriers Standards Association
EDFA	Erbium-Doped Fiber Amplifier
EDI	Electronic Data Interchange
EDIBANX	EDI Bank Alliance Network Exchange
EDIFACT	Electronic Data Interchange For Administration, Commerce, and Trade (ANSI)
EFCI	Explicit Forward Congestion Indicator
EFTA	European Free Trade Association
EGP	Exterior Gateway Protocol (IETF)
EIA	Electronics Industry Association
EIGRP	Enhanced Interior Gateway Routing Protocol
EIR	Excess Information Rate
EMBARC	Electronic Mail Broadcast to a Roaming Computer
EMI	Electromagnetic Interference
EMS	Element Management System
EN	End Node
ENIAC	Electronic Numerical Integrator and Computer
EO	End Office
EOC	Embedded Operations Channel (SONET)

EOT	End of Transmission (BISYNC)
EPROM	Erasable Programmable Read Only Memory
ESCON	Enterprise System Connection (IBM)
ESF	Extended Superframe Format
ESP	Enhanced Service Provider
ESS	Electronic Switching System
ETSI	European Telecommunications Standards Institute
ETX	End of Text (BISYNC)
EWOS	European Workshop for Open Systems
FACTR	Fujitsu Access and Transport System
FAQ	Frequently Asked Questions
FAT	File Allocation Table
FCS	Frame Check Sequence
FDD	Frequency Division Duplex
FDDI	Fiber Distributed Data Interface
FDM	Frequency Division Multiplexing
FDMA	Frequency Division Multiple Access
FDX	Full-Duplex
FEBE	Far End Block Error (SONET)
FEC	Forward Error Correction
FEC	Forward Equivalence Class
FECN	Forward Explicit Congestion Notification
FEP	Front-End Processor
FERF	Far End Receive Failure (SONET)
FET	Field Effect Transistor
FHSS	Frequency Hopping Spread Spectrum
FIB	Forward Indicator Bit (SS7)
FIFO	First In First Out
FITL	Fiber In The Loop

FLAG	Fiber Ling Across the Globe
FM	Frequency Modulation
FPGA	Field Programmable Gate Array
FR	Frame Relay
FRAD	Frame Relay Access Device
FRBS	Frame Relay Bearer Service
FSK	Frequency Shift Keying
FSN	Forward Sequence Number (SS7)
FTAM	File Transfer, Access, and Management
FTP	File Transfer Protocol (IETF)
FTTC	Fiber To The Curb
FTTH	Fiber To The Home
FUNI	Frame User-to-Network Interface
FWM	Four Wave Mixing
GATT	General Agreement on Tariffs and Trade
GbE	Gigabit Ethernet
Gbps	Gigabits per second (billion bits per second)
GDMO	Guidelines for the Development of Managed Objects
GEOS	Geosynchronous Earth Orbit Satellites
GFC	Generic Flow Control (ATM)
GFI	General Format Identifier (X.25)
GOSIP	Government Open Systems Interconnection Profile
GPS	Global Positioning System
GRIN	Graded Index (fiber)
GSM	Global System for Mobile Communications
GUI	Graphical User Interface
HDB3	High Density, Bipolar 3 (E-Carrier)
HDLC	High-level Data Link Control
HDSL	High-bit-rate Digital Subscriber Line

HDTV	High Definition Television
HDX	Half-Duplex
HEC	Header Error Control (ATM)
HFC	Hybrid Fiber/Coax
HFS	Hierarchical File Storage
HLR	Home Location Register
HSSI	High-Speed Serial Interface (ANSI)
HTML	Hypertext Markup Language
HTTP	Hypertext Transfer Protocol (IETF)
HTU	HDSL Transmission Unit
I	Intrapictures
IAB	Internet Architecture Board (formerly Internet Activities Board)
IACS	Integrated Access and Cross-connect System
IAD	Integrated Access Device
IAM	Initial Address Message (SS7)
IANA	Internet Address Naming Authority
ICMP	Internet Control Message Protocol (IETF)
IDP	Internet Datagram Protocol
IEC	Interexchange Carrier (also IXC)
IEC	International Electrotechnical Commission
IEEE	Institute of Electrical and Electronics Engineers
IETF	Internet Engineering Task Force
IFRB	International Frequency Registration Board
IGP	Interior Gateway Protocol (IETF)
IGRP	Interior Gateway Routing Protocol
ILEC	Incumbent Local ExchangeCarrier
IML	Initial Microcode Load
IMP	Interface Message Processor (ARPANET)
IMS	Information Management System

InARP	Inverse Address Resolution Protocol (IETF)
InATMARP	Inverse ATMARP
INMARSAT	International Maritime Satellite Organization
INP	Internet Nodal Processor
InterNIC	Internet Network Information Center
IP	Internet Protocol (IETF)
IPX	Internetwork Packet Exchange (NetWare)
ISDN	Integrated Services Digital Network
ISO	International Organization for Standardization
ISOC	Internet Society
ISP	Internet Service Provider
ISUP	ISDN User Part (SS7)
IT	Information Technology
ITU	International Telecommunication Union
ITU-R	International Telecommunication Union-Radio Communication Sector
IVD	Inside Vapor Deposition
IVR	Interactive Voice Response
IXC	Interexchange Carrier
JEPI	Joint Electronic Paynets Initiative
JES	Job Entry System
JIT	Just In Time
JPEG	Joint Photographic Experts Group
KB	Kilobytes
Kbps	Kilobits per second (thousand bits per second)
KLTN	Potassium Lithium Tantalate Niobate
LAN	Local Area Network
LANE	LAN Emulation
LAP	Link Access Procedure (X.25)
LAPB	Link Access Procedure Balanced (X.25)

LAPD	Link Access Procedure for the D-Channel
LAPF	Link Access Procedure to Frame Mode Bearer Services
LAPF-Core	Core Aspects of the Link Access Procedure to Frame Mode Bearer Services
LAPM	Link Access Procedure for Modems
LAPX	Link Access Procedure half-duplex
LASER	Light Amplification by the Stimulated Emission of Radiation
LATA	Local Access and Transport Area
LCD	Liquid Crystal Display
LCGN	Logical Channel Group Number
LCM	Line Concentrator Module
LCN	Local Communications Network
LD	Laser Diode
LDAP	Lightweight Directory Access Protocol (X.500)
LEAF®	Large Effective Area Fiber® (Corning product)
LEC	Local Exchange Carrier
LED	Light Emitting Diode
LENS	Lightwave Efficient Network Solution (Centerpoint)
LEOS	Low Earth Orbit Satellites
LI	Length Indicator
LIDB	Line Information Database
LIFO	Last In First Out
LIS	Logical IP Subnet
LLC	Logical Link Control
LMDS	Local Multipoint Distribution System
LMI	Local Management Interface
LMOS	Loop Maintenance Operations System
LORAN	Long-range Radio Navigation
LPC	Linear Predictive Coding
LPP	Lightweight Presentation Protocol

LRC	Longitudinal Redundancy Check (BISYNC)
LS	Link State
LSI	Large Scale Integration
LSP	Label Switched Path
LU	Line Unit
LU	Logical Unit (SNA)
MAC	Media Access Control
MAN	Metropolitan Area Network
MAP	Manufacturing Automation Protocol
MAU	Medium Attachment Unit (Ethernet)
MAU	Multistation Access Unit (Token Ring)
MB	Megabytes
MBA™	Metro Business Access™ (Ocular)
Mbps	Megabits per second (million bits per second)
MD	Message Digest (MD2, MD4, MD5) (IETF)
MDF	Main Distribution Frame
MEMS	Micro Electrical Mechanical System
MF	Multifrequency
MFJ	Modified Final Judgment
MHS	Message Handling System (X.400)
MIB	Management Information Base
MIC	Medium Interface Connector (FDDI)
MIME	Multipurpose Internet Mail Extensions (IETF)
MIPS	Millions of Instructions Per Second
MIS	Management Information Systems
MITI	Ministry of International Trade and Industry (Japan)
ML-PPP	Multilink Point-to-Point Protocol
MMDS	Multichannel, Multipoint Distribution System
MMF	Multimode Fiber

MNP	Microcom Networking Protocol
MP	Multilink PPP
MPEG	Motion Picture Experts Group
MPLS	Multiprotocol Label Switching
MPOA	Multiprotocol Over ATM
MRI	Magnetic Resonance Imaging
MSB	Most Significant Bit
MSC	Mobile Switching Center
MSO	Mobile Switching Office
MSVC	Meta-Signaling Virtual Channel
MTA	Major Trading Area
MTBF	Mean Time Between Failure
MTP	Message Transfer Part (SS7)
MTTR	Mean Time to Repair
MTU	Maximum Transmission Unit
MVS	Multiple Virtual Storage
NAFTA	North American Free Trade Agreement
NAK	Negative Acknowledgment (BISYNC, DDCMP)
NAP	Network Access Point (Internet)
NARUC	National Association of Regulatory Utility Commissioners
NASA	National Aeronautics and Space Administration
NASDAQ	National Association of Securities Dealers Automated Quotations
NATA	North American Telecommunications Association
NATO	North Atlantic Treaty Organization
NAU	Network Accessible Unit
NCP	Network Control Program
NCSA	National Center for Supercomputer Applications
NCTA	National Cable Television Association
NDIS	Network Driver Interface Specifications

NDSF	Non-Dispersion-Shifted Fiber
NetBEUI	NetBIOS Extended User Interface
NetBIOS	Network Basic Input/Output System
NFS	Network File System (Sun)
NIC	Network Interface Card
NII	National Information Infrastructure
NIST	National Institute of Standards and Technology (formerly NBS)
NIU	Network Interface Unit
NLPID	Network Layer Protocol Identifier
NLSP	NetWare Link Services Protocol
NM	Network Module
Nm	Nanometer
NMC	Network Management Center
NMS	Network Management System
NMT	Nordic Mobile Telephone
NMVT	Network Management Vector Transport protocol
NNI	Network Node Interface
NNI	Network-to-Network Interface
NOC	Network Operations Center
NOCC	Network Operations Control Center
NOS	Network Operating System
NPA	Numbering Plan Area
NREN	National Research and Education Network
NRZ	Non-Return to Zero
NRZI	Non-Return to Zero Inverted
NSA	National Security Agency
NSAP	Network Service Access Point
NSAPA	Network Service Access Point Address
NSF	National Science Foundation

NTSC	National Television Systems Committee
NTT	Nippon Telephone and Telegraph
NVOD	Near Video On Demand
NZDSF	Non-Zero Dispersion-Shifted Fiber
OADM	Optical Add-Drop Multiplexer
OAM	Operations, Administration, and Maintenance
OAM&P	Operations, Administration, Maintenance, and Provisioning
OAN	Optical Area Network
OC	Optical Carrier
OEM	Original Equipment Manufacturer
O-E-O	Optical-Electrical-Optical
OLS	Optical Line System (Lucent)
OMAP	Operations, Maintenance, and Administration Part (SS7)
ONA	Open Network Architecture
ONU	Optical Network Unit
OOF	Out of Frame
OS	Operating System
OSF	Open Software Foundation
OSI	Open Systems Interconnection (ISO, ITU-T)
OSI-RM	Open Systems Interconnection Reference Model
OSPF	Open Shortest Path First (IETF)
OSS	Operation Support Systems
OTDM	Optical Time Division Multiplexing
OTDR	Optical Time-Domain Reflectometer
OUI	Organizationally Unique Identifier (SNAP)
OVD	Outside Vapor Deposition
P/F	Poll/Final (HDLC)
PAD	Packet Assembler/Disassembler (X.25)
PAL	Phase Alternate Line

PAM	Pulse Amplitude Modulation
PANS	Pretty Amazing New Stuff
PBX	Private Branch Exchange
PCI	Pulse Code Modulation
PCMCIA	Personal Computer Memory Card International Association
PCN	Personal Communications Network
PCS	Personal Communications Services
PDA	Personal Digital Assistant
PDH	Plesiochronous Digital Hierarchy
PDU	Protocol Data Unit
PIN	Positive-Intrinsic-Negative
PING	Packet Internet Groper (TCP/IP)
PLCP	Physical Layer Convergence Protocol
PLP	Packet Layer Protocol (X.25)
PM	Phase Modulation
PMD	Physical Medium Dependent (FDDI)
PNNI	Private Network Node Interface (ATM)
PON	Passive Optical Networking
POP	Point of Presence
POSIT	Profiles for Open Systems Interworking Technologies
POSIX	Portable Operating System Interface for UNIX
POTS	Plain Old Telephone Service
PPP	Point-to-Point Protocol (IETF)
PRC	Primary Reference Clock
PRI	Primary Rate Interface
PROFS	Professional Office System
PROM	Programmable Read Only Memory
PSDN	Packet Switched Data Network
PSK	Phase Shift Keying

PSPDN	Packet Switched Public Data Network
PSTN	Public Switched Telephone Network
PTI	Payload Type Identifier (ATM)
PTT	Post, Telephone, and Telegraph
PU	Physical Unit (SNA)
PUC	Public Utility Commission
PVC	Permanent Virtual Circuit
QAM	Quadrature Amplitude Modulation
Q-bit	Qualified data bit (X.25)
QLLC	Qualified Logical Link Control (SNA)
QoS	Quality of Service
QPSK	Quadrature Phase Shift Keying
QPSX	Queued Packet Synchronous Exchange
R&D	Research & Development
RADSL	Rate Adaptive Digital Subscriber Line
RAID	Redundant Array of Inexpensive Disks
RAM	Random Access Memory
RARP	Reverse Address Resolution Protocol (IETF)
RAS	Remote Access Server
RBOC	Regional Bell Operating Company
RF	Radio Frequency
RFC	Request For Comments (IETF)
RFH	Remote Frame Handler (ISDN)
RFI	Radio Frequency Interference
RFP	Request For Proposal
RHC	Regional Holding Company
RHK	Ryan, Hankin, and Kent (Consultancy)
RIP	Routing Information Protocol (IETF)
RISC	Reduced Instruction Set Computer

RJE	Remote Job Entry
RNR	Receive Not Ready (HDLC)
ROM	Read-Only Memory
ROSE	Remote Operation Service Element
RPC	Remote Procedure Call
RR	Receive Ready (HDLC)
RTS	Request To Send (EIA-232-E)
S/DMS	SONET/Digital Multiplex System
S/N	Signal-to-Noise Ratio
SAA	Systems Application Architecture (IBM)
SAAL	Signaling ATM Adaptation Layer (ATM)
SABM	Set Asynchronous Balanced Mode (HDLC)
SABME	Set Asynchronous Balanced Mode Extended (HDLC)
SAC	Single Attachment Concentrator (FDDI)
SAN	Storage Area Network
SAP	Service Access Point (generic)
SAPI	Service Access Point Identifier (LAPD)
SAR	Segmentation And Reassembly (ATM)
SAS	Single Attachment Station (FDDI)
SASE	Specific Applications Service Element (subset of CASE, Application Layer)
SATAN	System Administrator Tool for Analyzing Networks
SBS	Stimulated Brillouin Scattering
SCCP	Signaling Connection Control Point (SS7)
SCP	Service Control Point (SS7)
SCREAM™	Scalable Control of a Rearrangeable Extensible Array of Mirrors™ (Calient)
SCSI	Small Computer Systems Interface
SCTE	Serial Clock Transmit External (EIA-232-E)
SDH	Synchronous Digital Hierarchy (ITU-T)

SDLC	Synchronous Data Link Control (IBM)
SDS	Scientific Data Systems
SECAM	Sequential Color with Memory
SF	Superframe Format (T-1)
SGML	Standard Generalized Markup Language
SGMP	Simple Gateway Management Protocol (IETF)
S-HTTP	Secure HTTP (IETF)
SIF	Signaling Information Field
SIG	Special Interest Group
SIO	Service Information Octet
SIR	Sustained Information Rate (SMDS)
SLA	Service Level Agreement
SLIP	Serial Line Interface Protocol (IETF)
SM	Switching Module
SMAP	System Management Application Part
SMDS	Switched Multimegabit Data Service
SMF	Single Mode Fiber
SMP	Simple Management Protocol
SMP	Switching Module Processor
SMR	Specialized Mobile Radio
SMS	Standard Management System (SS7)
SMTP	Simple Mail Transfer Protocol (IETF)
SNA	Systems Network Architecture (IBM)
SNAP	Subnetwork Access Protocol
SNI	Subscriber Network Interface (SMDS)
SNMP	Simple Network Management Protocol (IETF)
SNP	Sequence Number Protection
SONET	Synchronous Optical Network
SPAG	Standards Promotion and Application Group
SPARC	Scalable Performance Architecture

SPE	Synchronous Payload Envelope (SONET)
SPID	Service Profile Identifier (ISDN)
SPM	Self Phase Modulation
SPOC	Single Point of Contact
SPX	Sequenced Packet Exchange (NetWare)
SQL	Structured Query Language
SRB	Source Route Bridging
SRS	Stimulated Raman Scattering
SRT	Source Routing Transparent
SS7	Signaling System 7
SSL	Secure Socket Layer (IETF)
SSP	Service Switching Point (SS7)
SST	Spread Spectrum Transmission
STDM	Statistical Time Division Multiplexing
STM	Synchronous Transfer Mode
STM	Synchronous Transport Module (SDH)
STP	Signal Transfer Point (SS7)
STS	Synchronous Transport Signal (SONET)
STX	Start of Text (BISYNC)
SVC	Signaling Virtual Channel (ATM)
SVC	Switched Virtual Circuit
SXS	Step-by-Step Switching
SYN	Synchronization
SYNTRAN	Synchronous Transmission
TA	Terminal Adapter (ISDN)
TAG	Technical Advisory Group
TASI	Time Assigned Speech Interpolation
TAXI	Transparent Asynchronous Transmitter/Receiver Interface (Physical Layer)
TCAP	Transaction Capabilities Application Part (SS7)

TCM	Time Compression Multiplexing
TCM	Trellis Coding Modulation
TCP	Transmission Control Protocol (IETF)
TDD	Time Division Duplexing
TDM	Time Division Multiplexing
TDMA	Time Division Multiple Access
TDR	Time Domain Reflectometer
TE1	Terminal Equipment type 1 (ISDN capable)
TE2	Terminal Equipment type 2 (non-ISDN capable)
TEI	Terminal Endpoint Identifier (LAPD)
TELRIC	Total Element Long-Run Incremental Cost
TIA	Telecommunications Industry Association
TIRKS	Trunk Integrated Record Keeping System
TL1	Transaction Language 1
TM	Terminal Multiplexer
TMN	Telecommunications Management Network
TMS	Time-Multiplexed Switch
TOH	Transport Overhead (SONET)
TOP	Technical and Office Protocol
ToS	Type of Service (IP)
TP	Twisted Pair
TR	Token Ring
TRA	Traffic Routing Administration
TSI	Time Slot Interchange
TSLRIC	Total Service Long-Run Incremental Cost
TSO	Terminating Screening Office
TSO	Time-Sharing Option (IBM)
TSR	Terminate and Stay Resident
TSS	Telecommunication Standardization Sector (ITU-T)
TST	Time-Space-Time Switching

TSTS	Time-Space-Time-Space Switching
TTL	Time To Live
TU	Tributary Unit (SDH)
TUG	Tributary Unit Group (SDH)
TUP	Telephone User Part (SS7)
UA	Unnumbered Acknowledgment (HDLC)
UART	Universal Asynchronous Receiver Transmitter
UBR	Unspecified Bit Rate (ATM)
UDI	Unrestricted Digital Information (ISDN)
UDP	User Datagram Protocol (IETF)
UHF	Ultra High Frequency
UI	Unnumbered Information (HDLC)
UNI	User-to-Network Interface (ATM, FR)
UNIT™	Unified Network Interface Technology™ (Ocular)
UNMA	Unified Network Management Architecture
UPS	Uninterruptable Power Supply
UPSR	Unidirectional Path Switched Ring
UPT	Universal Personal Telecommunications
URL	Uniform Resource Locator
USART	Universal Synchronous Asynchronous Receiver Transmitter
UTC	Coordinated Universal Time
UTP	Unshielded Twisted Pair (Physical Layer)
UUCP	UNIX-UNIX Copy
VAN	Value-Added Network
VAX	Virtual Address Extension (DEC)
vBNS	Very High Speed Backbone Network Service
VBR	Variable Bit Rate (ATM)
VBR-NRT	Variable Bit Rate-Non-Real-Time (ATM)
VBR-RT	Variable Bit Rate-Real-Time (ATM)
VC	Virtual Channel (ATM)

VC	Virtual Circuit (PSN)
VC	Virtual Container (SDH)
VCC	Virtual Channel Connection (ATM)
VCI	Virtual Channel Identifier (ATM)
VCSEL	Vertical Cavity Surface Emitting Laser
VDSL	Very High-speed Digital Subscriber Line
VDSL	Very High bit rate Digital Subscriber Line
VERONICA	Very Easy Rodent-Oriented Netwide Index to Computerized Archives (Internet)
VGA	Variable Graphics Array
VHF	Very High Frequency
VHS	Video Home System
VINES	Virtual Networking System (Banyan)
VIP	VINES Internet Protocol
VLF	Very Low Frequency
VLR	Visitor Location Register (Wireless/GSM)
VLSI	Very Large Scale Integration
VM	Virtual Machine (IBM)
VM	Virtual Memory
VMS	Virtual Memory System (DEC)
VOD	Video-On-Demand
VP	Virtual Path
VPC	Virtual Path Connection
VPI	Virtual Path Identifier
VPN	Virtual Private Network
VR	Virtual Reality
VSAT	Very Small Aperture Terminal
VSB	Vestigial Sideband
VSELP	Vector-Sum Excited Linear Prediction
VT	Virtual Tributary

VTAM	Virtual Telecommunications Access Method (SNA)
VTOA	Voice and Telephony Over ATM
VTP	Virtual Terminal Protocol (ISO)
WACK	Wait Acknowledgment (BISYNC)
WACS	Wireless Access Communications System
WAIS	Wide Area Information Server (IETF)
WAN	Wide Area Network
WARC	World Administrative Radio Conference
WATS	Wide Area Telecommunications Service
WDM	Wavelength Division Multiplexing
WIN	Wireless In-building Network
WTO	World Trade Organization
WWW	World Wide Web (IETF)
WYSIWYG	What You See Is What You Get
xDSL	x-Type Digital Subscriber Line
XID	Exchange Identification (HDLC)
XNS	Xerox Network Systems
XPM	Cross Phase Modulation
ZBTSI	Zero Byte Time Slot Interchange
ZCS	Zero Code Suppression

INDEX